British
Wildlife

○ Collins **wild guide**

British
Wildlife

HarperCollinsPublishers Ltd.
77-85 Fulham Palace Road
London
W6 8JB

The Collins website address is:
www.collins.co.uk

Collins is a registered trademark of
HarperCollinsPublishers Ltd.

First published in this edition in 2002

Text and pictures are taken from the following Wild Guide titles:
Birds by Peter Holden
Wild Flowers by John Akeroyd
Insects by Bob Gibbons
Mushrooms by Brian Spooner
Wild Animals by John Burton
Trees by Bob Press
Butterflies and Moths by John Still

10 09 08 07 06 05

10 9 8 7 6 5 4 3 2 1

A catalogue record for this book is available from the British Library.

ISBN 0-00-719172-3

Colour reproduction by Colourscan, Singapore
Printed and bound by Printing Express Ltd, Hong Kong

CONTENTS

HOW TO USE THIS BOOK

This book is a basic introduction to many species of wildlife you are likely to encounter in the British Isles, and many will also be found in the other countries bordering the North Sea. The book is intended to be taken out into the field, either in a pocket or on the dashboard or parcel shelf of a car. Its first purpose is to help you identify what you see, and tell you more about it.

Nearly 500 species are described and illustrated with a **colour photograph**. The **main text** gives an introduction to the species, its behaviour and range. The **ID Fact File** gives information at a glance. Where relevant, a **calendar bar** gives a month-by month guide to when you can expect to see the species. There may also be a map or additional illustrations.

Birds

In this section, as well as the photograph, illustrations show other common plumages, such as the female, a young bird or the winter plumage. So if the photograph does not quite match what you are looking at, remember to check the illustrations as well. Maps show the bird's distribution in Britain and the rest of Europe. The red areas represent the bird's breeding range; the area below the heavy dotted line, or enclosed by it, is the bird's winter range. This does not mean that the bird occurs everywhere within these limits, but locally where its proper habitat is available. If the bird's winter range is identical to its breeding range, or if the bird leaves the area entirely in winter there is no dotted line. Dark green in the calendar bar indicates the months in which the bird definitely occurs in Britain; mid-green means it may occur; light green is used when it does not occur. The coloured icon in the top left-hand corner represents the family to which the bird belongs.

Wild Flowers

Flowers are grouped together by family, but bear in mind that some, such as the Rose and Figwort families, display a range of flower structures that can confuse even the most experienced botanist. The coloured symbol gives a thumbnail sketch of the shape and colour of the flower or flower structure. Here the map shows the distribution of each plant in western Europe. It indicates parts of Britain and Ireland from which the plant is absent and also shows how some plants reach the edge of their distribution here. The shaded sections of the calendar bar show the months in which each species is most likely to flower. Some wild flowers, like orchids, have short, precise flowering periods. Others like the Red Campion, have a distinct flowering season but may be found in flower at almost any time. Some annual weeds, such as Chickweed and Groundsel, flower throughout the year except during severe frost and snow.

Trees
Some of the more common native and introduced tree species found in Britain are included. They are arranged in order of their closest relationships, and the leaf symbols indicates the family to which the tree belongs. It will help you to see which trees are related to each other, and can be used as a quick guide when identifying a tree since the general shape is typical for each family (although there are exceptions). So when identifying a tree, look for the leaf symbol that most closely matches the one you are looking at, and the species you want should be nearby. The lighter portion of the calendar bar indicates those months in which you will usually find the tree in flower. For conifers the lighter portion indicates when the tree is likely to be shedding pollen from male cones. The illustrations show additional details such as winter silhouettes, fruit, flowers and leaves.

Mammals
The more common land and marine mammals are covered in this section. The symbols indicate the group which they belong, and the shaded part of the calendar bar shows when they might be seen.

Insects
A coloured symbol represents the different orders or groups of insects, and this can act as an aid to quick reference.

The calendar bar is shaded during the months when the insects is most likely to be seen, but these will vary according to location and season.

Mushrooms and Toadstools

The symbols in this section (see below) indicate the practical grouping to which the mushrooms belong.

The calendar bar shows the months in which the fungus fruitbody is usually found.

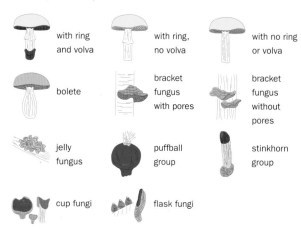

with ring and volva

with ring, no volva

with no ring or volva

bolete

bracket fungus with pores

bracket fungus without pores

jelly fungus

puffball group

stinkhorn group

cup fungi

flask fungi

Butterflies and Moths

Here the symbols represent the general shape and colour of the family to which the species belongs. The symbol of the sun or the moon indicates whether the insect flies by day or night, or in some cases, both. If the female is very different from the male, or identification is helped by looking at the underside (the view of the butterfly with its wings closed), this is shown by the illustration. The illustrations mainly show both upperside and underside (undersides are marked with a grey triangle). If the sexes are very similar then the caterpillar is shown. The calendar bar gives an indication of when you can expect to see the adult insects.

Amphibians and Reptiles

Some common amphibians and reptiles are covered in this section. The symbols indicate the group which they belong, and the shaded part of the calendar bar shows when they might be seen.

Fish

Some common freshwater fish are featured in this last section. The shaded part of the calendar bar shows when they might be seen.

WHERE TO WATCH WILDLIFE

Parks and Gardens

There are a number of birds that form part of the urban environment in the British Isles, such as Feral Pigeons, House Sparrows, Starlings and Swifts. Sparrowhawks are becoming more common, and are a threat to smaller songbirds. Parks are home to a greater variety of species, many associated with woodland. Some mammals, such as foxes and mice, thrive in towns and cities, and others such as hedgehogs are also seen. Mature gardens and flowers in parks will attract butterflies, but you will not find many different species as only a few have adjusted to the change from countryside to towns. You will notice that butterflies are attracted to certain type of flowers. In man-made habitats the soil is often nitrogen-rich due to the use of fertilisers. These encourage the growth of a range of fungi, especially the true mushrooms (*Agaricus* species), some of the Ink-caps and Cone-heads, Parasol Mushrooms and Hay-caps amongst many others. Many other species will be found on old rotting stumps and trunks, and many of the grassland species will also occur.

Meadows, Grassland and Farmland

Butterflies are in greater evidence in grassland habitats. Most of the smaller butterflies, like many of the blues and skippers, live in small colonies and never fly very far from the area in which they were born. Other species however, such as the Painted Lady, fly long distances. Grassland is a man-made habitat and the British Isles at the end of the 20th Century is one of the world's most modified landscapes. Grasslands have been colonized by flower species displaced by woodland, and a flower-rich meadow is one of the most interesting habitats to visit. The species found there depend on whether they favour dry or damp conditions, and the type of underlying soil. The most interesting, and often now the most uncommon, of the grassland fungi are those which occur in unimproved, nutrient-poor meadows. This type of habitat supports a rich community of wild flowers and fungi such as the attractive, bright-coloured Wax-caps, some Fairy-club fungi and Earth-tongues. Regrettably, these habitats are becoming scarce and as a result many of these fungi are now much less common.

Woodlands

Britain used to be covered in coniferous and deciduous woods, but man's impact on the landscape means that now small pockets of woodland, and some larger areas remain. Deciduous woodland is rich in all types of wildlife: birds, mammals, insects, butterflies and wild flowers, as well as our native tree species. Usually the soil type in each area determines the dominant species in any woodland. Resident woodland birds include tits, Chaffinches, woodpeckers and owls. Other species are migrants which arrive in the spring, such as warblers and flycatchers. Woodlands form a welcome refuge for mammals which have been driven from intensively farmed areas. Amongst the richest habitats for larger fungi, both deciduous and coniferous woods are ideal places to look for a wide range of species. In general, species found in deciduous woods differ from those found with conifers, and many of the woodland toadstools grow only with certain kinds of tree.

Heathlands and Dunes

A very specialised habitat, British heathlands are under threat. Soils are generally acid and support a variety of species of heather and gorse. Where water collects bogs and marshes are formed which are home to other plants such as mosses. Heathlands are rich in bird species, and reptiles such as Adder and Common Lizard. Moths and butterflies can be seen, and a variety of insects inhabit areas near water. In undisturbed areas, sand dunes support a variety of flower and grass species; colonies of sea birds such as waders, gulls and terns may also be found. Coastal dunes support a distinct group of fungal species, most of them adapted to grow in a particular part of the dune system. The fungi of waterlogged habitats will vary according to the amount of tree cover and whether the soil is acid or alkaline.

Estuary and Saltmarsh

The brackish water found in these apparently inhospitable areas is home to particular types of plants such as Sea Lavender. They are also rich in birdlife from September to March, including ducks, geese and waders.

Freshwater Habitats

Fast-flowing or still water habitats are good places to watch wildlife. Water supports rich vegetation, again the species will depend on the soil type; for exmaple, watercress is found in lowland chalk streams. Fish, insects, birds and mammals live in and around the water. Swans, ducks and Kingfishers will swim or feed in the water. Insect eating species`are attracted by the flies that skim over the surface. In still water look for frogs and toads when they congregate for the mating season in early spring. In the autumn wildfowl gather together for the winter.

FURTHER READING

If you want to take your interest in any or all of the subjects featured in this book further, Collins Natural History offers a range of guides.

Collins Gem Guides
Handy-sized beginners' guides, full of colour photographs and identification information.
Birds
Butterflies and Moths
Insects
Mushrooms and Toadstools
Seashore
Trees
Wild Flowers

Collins How To Identify Guides
Illustrated with original artwork, these innovative guides are designed to help those with the most basic knowledge identify different types of wildlife and phenomena with confidence.
Butterflies by Richard Lewington
Edible Mushrooms by Patrick Harding, Tony Lyon and Gill Tomblin
Trees by Patrick Harding and Gill Tomblin
Wild Flowers by Christopher Grey-Wilson and Lisa Alderson

Other titles
Collins Guide to Birds of Britain and Ireland by Dominic Couzens
Complete British Wildlife by Paul Sterry
Garden Bird Songs and Calls by Geoff Sample
Gardening for Birds by Stephen Moss

27 cm

J	F	M	A	M	J
J	A	S	O	N	D

Little Grebe
Tachybaptus ruficollis

ID FACT FILE

SIZE: Smaller than Moorhen

SUMMER: Dark brown body, paler underparts, reddish-brown face and neck

WINTER: Paler brown and grey

YOUNG: Brown streaks on head

BILL: Short and pointed, yellow base in summer

IN FLIGHT: Reluctant to fly. No white patches in wings

VOICE: Loud trill

LOOKALIKES: Black-necked Grebe, Slavonian Grebe

Small and secretive, the Little Grebe breeds on lakes, quiet rivers and small ponds in central and S Europe. There is some migration in spring and autumn. Sometimes the birds form flocks in winter. Dives when disturbed and when feeding. Food is insects and small fish. A floating nest of water weed is built among vegetation, and the 5 eggs hatch after 20 days. Young are cared for by both parents and often ride on their parents' backs. They fly after 46 days. There are 2 broods.

adult summer

adult winter

46–51 cm

J	F	M	A	M	J
J	A	S	O	N	D

ID FACT FILE

Size: Smaller than Mallard

Summer: Long white neck and white underparts, brown back, orange-brown crest and ear-tufts

Winter: Grey and white

Juvenile: Like winter adult. Stripy head and neck

Bill: Dagger-like

In flight: White patches on wings. Neck held out straight in front. Trailing feet

Voice: Low, growling

Lookalikes: Red-necked Grebe, Red-throated Diver

Great Crested Grebe
Podiceps cristatus

Breeds on inland waters, but may be seen on the sea in winter and sometimes forms flocks. Northern grebes migrate south or west in autumn. Dives to find fish, and has elaborate courtship display. A floating nest is made of aquatic vegetation attached to water plants. The 3–5 eggs are incubated by both adults for 28 days. Young swim soon after hatching and ride on their parents' backs to protect them from predators such as pike. They fly after 71 days. There are 1 or 2 broods.

adult summer

adult winter

80–100 cm

J	F	M	A	M	J
J	A	S	O	N	D

Cormorant
Phalacrocorax carbo

ID FACT FILE

Size: Like a large goose

Breeding: Black, pale throat, white on back of head and on thighs

Non-breeding: Black, pale throat

Juvenile: Dark brown upperparts, paler underparts darkening over first 2 years

Bill: Long, strong with hooked tip

In flight: Goose-like with slow, strong wing-beats, neck held out in front

Voice: Deep, guttural calls when nesting

Lookalikes: Shag, divers

Cormorants frequently stand with their wings partly spread. This is a mainly coastal bird, but it is also found along river valleys and on lakes far from the sea. It dives to feed on fish. Cormorants nest in colonies on cliff ledges, or sometimes in trees. The nest is made of twigs and seaweed. The 3–4 eggs hatch after 30 days and the young leave after 50 days, but may return to be fed for a further 50 days.

adult non-breeding

juvenile

immature

70–80 cm

J	F	M	A	M	J
J	A	S	O	N	D

Bittern
Botaurus stellaris

ID FACT FILE

Size: Smaller than Grey Heron

All birds: Stocky, thick-necked, golden-brown with black streaks and black crown

Bill: Long and dagger-like

In flight: Broad, rounded wings. Neck hunched up and legs trailing

Voice: Deep booming call of male in spring

Lookalikes: Grey Heron

Secretive and solitary, this is a well-camouflaged bird of lowland marshes and dense reed-beds. It feeds mainly on fish, especially eels, but also eats other animals including small birds. Some northern Bitterns migrate in autumn. Males may have several mates. The nest is a heap of dead reeds. The 4–6 eggs hatch after 25 days. Young may leave the nest after 15 days and fly after 50 days.

adult

adult

90–98 cm

J	F	M	A	M	J
J	A	S	O	N	D

Grey Heron
Ardea cinerea

ID FACT FILE

SIZE: Large, long necked, long-legged

ADULT: Grey and white with wispy black crest

JUVENILE: Greyer than adult and lacks crest

BILL: Long, dagger-like

IN FLIGHT: Slow, powerful wing-beats. Neck is drawn up on to shoulders, feet project beyond tail

VOICE: Harsh *frrank*

LOOKALIKES: None in area

This distinctive, large bird often stands beside water ready to grasp a passing fish, but may stand hunched up in a field away from water. It also feeds on amphibians or small mammals. Many European herons move south in autumn. Nests are in colonies (heronries) at the tops of tall trees. The nest is a large platform of sticks, in which 3–5 eggs are laid in February-April, hatching after 25 days. Young fly after 50 days and soon leave the area.

adult

immature

adult

87–100 cm

| J | F | M | A | M | J |
| J | A | S | O | N | D |

Gannet

Sula bassana

ID FACT FILE

Size: Larger than any gull

Adult: White, black wing-tips, yellow nape

Juvenile: Grey, gradually becoming white over 5 years

Bill: Dagger-like

In flight: Cigar-shaped with long, narrow, black-tipped wings

Voice: Usually silent, growling *urr* when nesting

Lookalikes: Skuas, Gulls and Terns

Birds of the open ocean, Gannets breed on small islands off the NW coast of Europe. They move away from land after nesting to winter at sea. The young migrate south as far as W Africa. Gannets feed on fish by plunge-diving from 25 m, and nest in large, noisy colonies. The nest is a pile of seaweed. A single egg is incubated for 44 days. The young bird is fed by both parents and flies after 90 days.

adults at breeding colony

adult

three immature stages

HERONS AND ALLIES

80–90 cm

J	F	M	A	M	J
J	A	S	O	N	D

Spoonbill
Platalea leucorodia

ID FACT FILE

SIZE: Smaller than Grey Heron

ADULT: White, short crest, and yellowish band on breast when breeding

JUVENILE: Black wing-tips. Lacks crest and yellow mark

BILL: Long and spoon-shaped

IN FLIGHT: Neck held straight out in front, feet trailing

VOICE: Usually silent. Bill snapping some-times at nest

LOOKALIKES: None in area

Coastal marshes and reed-beds in river valleys, mainly in SE Europe, are home for colonies of Spoonbills. They feed together in groups in open water, sweeping their bills from side to side as they filter the tiny creatures on which they feed. Most migrate away from colonies after nesting. The nest is a pile of reeds and twigs on the ground or in low trees or bushes. The 3–4 eggs are incubated for 24 days, and the young fly after about 24 days.

adult

adult

immature

145–160 cm

J	F	M	A	M	J
J	A	S	O	N	D

Whooper Swan

Cygnus cygnus

ID FACT FILE

Size: As Mute Swan

Adult: White, sometimes with rust-coloured stains on neck. Base of long, straight neck often rests on back

Juvenile: Brown with pink and black bill

Bill: Black with 'wedge' of yellow

In flight: Neck held straight

Voice: Trumpet-like call

Lookalikes: Mute Swan, Bewick's Swan

This swan breeds in Iceland and parts of N Europe. It migrates southwards in autumn to lakes, rivers and sheltered coastal inlets. Family parties usually stay together and large flocks form where food is plentiful. Feeds mainly on aquatic vegetation, but also grazes on arable fields in winter. The nest is a large mound of vegetation into which 3–5 eggs are laid. Young hatch after 35 days and fly about 80 days later.

adult

juvenile

adult

145–160 cm

| J | F | M | A | M | J |
| J | A | S | O | N | D |

Mute Swan
Cygnus olor

A familiar bird of lakes and slow-flowing rivers, mainly in NW Europe. Most are resident, but north-easterly populations migrate south and west in autumn. Flocks of non-breeding, mainly young, birds may gather. Pairs nest singly. The large nest is made of rushes or reeds. The 5–8 eggs hatch after 36 days. Young swim soon after hatching and may ride on their parents' backs. Immatures may stay with the parents during their first winter.

ID FACT FILE

Size: Very large with long neck

Male: All white with prominent black knob at base of bill

Female: As male, with smaller black knob

Immature: Greyish-brown, with more white showing as it grows older

Bill: Orange-red with black marks

In flight: Neck held out in front. Wings make loud rhythmic whistling

Voice: Snorts and hisses

Lookalikes: Whooper Swan, Bewick's Swan

adult

immature

adult

SWANS, GEESE AND DUCKS

65–78 cm

| J | F | M | A | M | J |
| J | A | S | O | N | D |

White-fronted Goose
Anser albifrons

ID FACT FILE

Size: Smaller than Greylag Goose

Adult: Deep chest and squarish head. Grey-brown, white forehead; black bars on belly

Juvenile: No white on face or black bars on belly

Bill: Orange or pink

In flight: Squarish head, deep chest, often very agile

Voice: More high-pitched than other grey geese

Lookalikes: Bean Goose, Pink-footed Goose, Greylag Goose

In winter this is the most numerous goose in Europe. It breeds on the Arctic tundra and migrates in autumn. Most European migrants come from Arctic Russia but some from Greenland winter in the British Isles, especially in Ireland. Feeds on leaves, roots and seeds. Young stay with their parents for the first autumn and winter and families often start the return migration together.

adult

juvenile

adult

75–90 cm

Greylag Goose
Anser anser

ID FACT FILE

Size: The largest brown goose

All birds: Brown, with thick, pale neck and large, pale head

Bill: Strong, triangular. Orange on western birds, pink on eastern populations

In flight: Heavy-looking, with large head, thick neck, blue-grey patches on wings

Voice: Deep honking calls

Lookalikes: Bean Goose, Pink-footed Goose, White-fronted Goose

The Greylag is the ancestor of many domesticated geese. It breeds in N and E European wetlands, from Arctic tundra to reed-beds. In autumn it migrates south and west to traditional wintering areas. It feeds on plant material, and lays 4–6 eggs in a nest of locally gathered vegetation. Young hatch after 27 days, fly about 50 days later and stay with their parents for the first autumn and winter.

adult

juvenile

adult

SWANS, GEESE AND DUCKS

58–70 cm

| J | F | M | A | M | J |
| J | A | S | O | N | D |

Barnacle Goose

Branta leucopsis

ID FACT FILE

Size: Smaller than Canada Goose

Adult: Black and grey with white face and barred back

Juvenile: Similar to adult but duller, with some brown feathers and a black line through the eye

Bill: Small and black

In flight: Looks black and white, with a longish neck and rather pointed wings

Voice: Dog-like yelping

Lookalikes: Canada Goose, Brent Goose

Barnacle Geese breed in the Arctic on steep cliffs and rock pinnacles or hummocks on lower ground. There are 3 separate populations, in Greenland, Spitsbergen and Novaya Zemlya, and all have separate migration routes and different wintering areas: the Novaya Zemlya population winters in the Netherlands, the others in the British Isles. They eat grasses and other vegetation. Young stay with their parents for the first autumn and winter.

adult

juvenile

adult

SWANS, GEESE AND DUCKS

51–61 cm

Brent Goose
Branta bernicla

ID FACT FILE

Size: Small, Mallard-sized goose

Adult: Small black head, white patch on side of black neck, dark brown body.

Siberian birds: Dark belly

Greenland birds: Pale belly

Juvenile: Like adult without neck patch, and with more barring on back

Bill: Short and black

In flight: Fast, with rather pointed wings and short neck

Voice: Flocks have murmuring, growling calls

Lookalikes: Barnacle Goose

These geese breed on the low tundra of the Arctic. Breeding must take place within 100 days before snow and ice return. Two populations winter in Europe: Siberian birds migrate across NW Europe to winter in the Netherlands, France and England; others from Greenland fly to Ireland with a few reaching Britain and France. In winter they feed on vegetation growing around coast and in estuaries, but also in arable fields near the sea.

adult dark-bellied

juvenile dark-bellied

adult light-bellied

SWANS, GEESE AND DUCKS

63–73 cm

J	F	M	A	M	J
J	A	S	O	N	D

Egyptian Goose
Alopochen aegyptiacus

ID FACT FILE

SIZE: Larger than Shelduck

ADULT: Long neck and legs. Grey-brown with darker back, dark patch round eye, dark spot on breast

JUVENILE: Darker head. Lacks dark patches round eye and on breast

BILL: Heavy-looking, pink and black

IN FLIGHT: Large white patches on wings

VOICE: Husky puffing and louder trumpeting

LOOKALIKES: Grey geese, Shelduck

Introduced to E England from Africa, these geese now breed wild in parkland and near some lowland lakes. Feeds mainly on leaves, grasses and seeds. Nests on a mound of leaves and reeds, laying 8 or 9 eggs which hatch after 28 days. Young feed themselves and are cared for by both parents. They fly after about 75 days.

adult

juvenile

adult

SWANS, GEESE AND DUCKS

58–67 cm

| J | F | M | A | M | J |
| J | A | S | O | N | D |

Shelduck
Tadorna tadorna

ID FACT FILE

Size: Larger than Mallard

Adult: Black and white with dark green head and orange band on breast

Juvenile: Grey-brown above, white below and pale face

Bill: Red. Male's has enlarged knob at base

In flight: Black and white, rather goose-like

Voice: Growling *ark-ark-ark*

Lookalikes: Shoveler, Eider, Goosander, Red-breasted Merganser

This is mainly a coastal species but sometimes breeds inland. Most live on estuaries or muddy shores where they filter molluscs, crustaceans and other invertebrates from the mud. Nests in enclosed sites such as rabbit burrows. Lays 8–10 eggs, which hatch after 29 days. The young join other broods and parents often leave them with an 'auntie' and migrate to a special moulting area. Young are independent at 20 days and able to fly after 45 days.

adult male

juvenile   adult

SWANS, GEESE AND DUCKS

45–51 cm

| J | F | M | A | M | J |
| J | A | S | O | N | D |

Wigeon
Anas penelope

ID FACT FILE

SIZE: Smaller than Mallard

MALE: Chestnut head with yellow crown. Grey body with pale pink breast, white wing-patches

FEMALE: Mottled brown, rounded crown

ECLIPSE: Jun–Oct. Male like female with white wing-patches

JUVENILE: First-winter males lack wing-patches

BILL: Rather small

IN FLIGHT: Pale belly, long, narrow wings

VOICE: Loud *we-ooo*

LOOKALIKES: Pochard, Teal, Gadwall, Mallard

The Wigeon is more likely than other ducks to be seen out of the water grazing on grass. A winter visitor to most of Europe, mainly to coastal areas, it will also follow river valleys to inland lakes and flooded fields. It feeds on leaves, stems and seeds, and nests among vegetation near lakes in N Europe. The 8 or 9 eggs hatch after 24 days. Young feed themselves and fly after about 40 days.

adult male

adult female

adult female

adult male

SWANS, GEESE AND DUCKS

46–56 cm

J	F	M	A	M	J
J	A	S	O	N	D

Gadwall
Anas strepera

ID FACT FILE

Size: Smaller than Mallard

Male: Finely marked grey body, black tail

Female: Mallard-like but more graceful, with white wing-patch

Eclipse: May–Sep. Male like female but a little greyer

Bill: Male's is grey, female's has orange sides

In flight: White wing-patch on both male and female

Voice: Male has croaking call, female a quiet quack

Lookalikes: Mallard, Pintail, Pochard

A delicately marked duck which breeds in scattered locations in many parts of Europe and winters in others. Feeds mainly on water plants. Nests close to lowland lakes with plenty of vegetation. Moves to larger lakes in winter and is joined by migrants from the north-east. The 8–12 eggs hatch after 24 days. Young are looked after by the female and can feed themselves. They fly after 45 days.

adult male

adult female

adult female

adult male

SWANS, GEESE AND DUCKS

34–38 cm

J	F	M	A	M	J
J	A	S	O	N	D

Teal
Anas crecca

ID FACT FILE

Size: Smaller than Mallard

Male: Chestnut and green head, white stripe along grey body, yellow patch under tail

Female: Like small female Mallard with green speculum

Eclipse: Jun–Sep. Male resembles dark female

Bill: Small and grey

In flight: Rapid, often twisting like waders

Voice: Piping

Lookalikes: Garganey

The smallest and most secretive European duck breeds in wetlands throughout N Europe and in a few southerly locations. Autumn migration takes it to many more freshwater and coastal marshes for the winter. Feeds on seeds and small creatures which it filters from the water or from soft mud. Nests under cover near water. Egg-laying starts in mid-April, and 8–11 eggs hatch after 21 days. Young feed themselves and fly after about 25 days.

adult male

adult female

adult female

adult male

50–65 cm

| J | F | M | A | M | J |
| J | A | S | O | N | D |

Mallard
Anas platyrhynchos

ID FACT FILE

SIZE: The largest common duck

MALE: Green head, grey body, purple-brown breast

FEMALE: Mottled brown

ECLIPSE: Jul–Sep. Male like female with dark crown and yellow bill

BILL: Large. Male's yellow, female's horn-coloured

IN FLIGHT: Blue speculum

VOICE: Female quacks, male has quieter *arrk*

LOOKALIKES: Other ducks

The most widespread duck in the world and familiar on park lakes as well as in remote wetlands throughout Europe. Northern populations migrate south and west in winter. Feeds on a variety of animal and vegetable material. Usually nests under cover, but may nest off the ground in trees. Egg-laying starts in March, and 9–13 eggs take 27 days to hatch. Young feed themselves and dive to escape danger. They fly after about 50 days.

adult male

adult female

adult female

adult male

51–66 cm

Pintail
Anas acuta

ID FACT FILE

Size: Mallard-sized with longer neck

Male: Grey body, long pointed tail, creamy breast and brown head with white stripe on neck

Female: Mallard-like, greyer, with longer neck

Eclipse: Jul–Oct. Male like female, but with darker upperparts

Bill: Grey

In flight: Long neck and tail

Voice: Like quiet Mallard

Lookalikes: Other dabbling ducks

This elegant, long-necked duck breeds near open, fresh water in N and E Europe, and sporadically farther south. In autumn migrants fly south and west mainly to coastal wetlands, especially estuaries. Pintails feed on a variety of plant and animal matter. Nests may be several hundred kilometres from water among grasses or rushes, and 7–9 eggs, laid in April and May, hatch after 22 days. Young feed themselves and fly after about 40 days.

adult male

adult female

adult female

adult male

SWANS, GEESE AND DUCKS

44–52 cm

J	F	M	A	M	J
J	A	S	O	N	D

Shoveler
Anas clypeata

ID FACT FILE

SIZE: Similar to Mallard

MALE: Green head, white breast and orange sides

FEMALE: Mallard-like, with green speculum

ECLIPSE: May–Sep, male is like female; Sep–Nov, male has scaly-looking breeding lumage

BILL: Very large and broad

IN FLIGHT: Wings set far back; blue forewings

VOICE: Hoarse *took, took*

LOOKALIKES: Mallard

The Shoveler uses its large, broad bill to filter small creatures and seeds from the water and mud. Shovelers nest near shallow inland waters with lots of vegetation. In winter they migrate south to similar habitats. In the British Isles most Shovelers migrate south, but are replaced by migrants from the north-east. The 9–11 eggs hatch after 22 days. Young are cared for by the female, feed themselves and fly after 40 days.

adult male

adult female

adult female

adult male

42–49 cm

SWANS, GEESE AND DUCKS

Pochard
Aythya ferina

ID FACT FILE

SIZE: Smaller than Mallard

MALE: Reddish-brown head, grey body, black breast and tail

FEMALE: Grey-brown body, blotchy cheeks

ECLIPSE: Jul–Aug. Male resembles female but greyer

BILL: Grey with silvery band

IN FLIGHT: Pale body contrasts with black breast and tail

VOICE: Usually silent, but wings make whistling sound

LOOKALIKES: Wigeon, Tufted Duck

A diving duck which breeds around inland lakes in many parts of Europe and reaches others as a winter migrant, when it often gathers in large flocks. It mostly feeds on plants that grow on the bottom of shallow lakes, which it reaches by diving. It nests in dense cover close to water. The 8–10 eggs hatch after 25 days. Young are cared for by the female, feed themselves and fly after 50 days.

adult male

adult female

adult female

adult male

40–47 cm

| J | F | M | A | M | J |
| J | A | S | O | N | D |

Tufted Duck
Aythya fuligula

ID FACT FILE

Size: Smaller than Mallard

Male: Black with white sides. Black drooping crest

Female: Dark brown with pale brown sides

Eclipse: Jul–Oct. Male resembles female with darker upper-parts and paler under-parts

Bill: Grey

In flight: White wing-bar, white belly on male

Voice: Variety of quiet calls

Lookalikes: Scaup, Golden-eye, Common Scoter

This diving duck breeds across N Europe and in some southerly locations. In autumn it migrates south and west to both inland and coastal areas. It feeds on water plants, invertebrates and especially freshwater mussels, diving up to 14 m to take food from the bottom of lakes. It nests near water and lays 8–11 eggs. Young hatch after 25 days and feed on midge larvae. They fly after 45 days.

adult male

adult female

adult female

adult male

50–71 cm

J	F	M	A	M	J
J	A	S	O	N	D

Eider
Somateria mollissima

ID FACT FILE

Size: As Mallard

Male: Black and white, with lime-green nape

Female: Dark, mottled brown

Eclipse: Jul–Sep. Male sooty-brown; forewings white

Juvenile: Young males have a variety of black-and-white plumages

Bill: Large, wedge-shaped

In flight: Male black behind, white in front; female brown and heavy-looking

Voice: Cooing calls in spring

Lookalikes: Long-tailed Duck, Goldeneye, Goosander, Shelduck

This marine diving duck breeds around rocky coasts of N Europe and winters only a little farther south. It tears mussels from rocks with its strong bill. The nest of down from the female's breast may be in the open or sheltered by rocks. The 4–6 eggs hatch after 25 days. Young feed themselves. While parents moult, creches of young are cared for by 'aunties'. Young fly after 65 days.

adult male

adult female

adult male

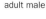

adult female

adult male eclipse

42–50 cm

Goldeneye
Bucephala clangula

ID FACT FILE

SIZE: Smaller than Mallard

MALE: White with dark head and back. White spot in front of eye

ECLIPSE: Aug–Sep, Male resembles female

FEMALE: Grey with brown head, white collar, white in wings

BILL: Rather small and dark

IN FLIGHT: Wings whistle. White wing-patches

VOICE: Usually silent. Low growls during courtship

LOOKALIKES: Long-tailed Duck, Smew, Eider, Tufted Duck

A large-headed diving duck which nests in northern coniferous forests where there are lakes or rivers. Nests in holes in trees and will use special nestboxes. Migrates south and west in winter and visits inland lakes and the coast. Feeds on molluscs, crustaceans and insect larvae. The 8–11 eggs hatch after 29 days. Young feed themselves and are brooded by female. They fly after about 57 days.

adult male

adult female

adult male

adult female

38–44 cm

J	F	M	A	M	J
J	A	S	O	N	D

Smew
Mergus albellus

ID FACT FILE

SIZE: Smaller than Mallard

MALE: White with black marks. Small crest

FEMALE AND JUVENILE: Grey. Reddish-brown head, white throat and neck

BILL: Small, dark grey

IN FLIGHT: Black and white wing-pattern. Male has white body

VOICE: Usually silent

LOOKALIKES: Goldeneye, Goosander, Red-breasted Merganser

This smallest member of the Merganser family is an attractive diving duck and a winter visitor from northern forests to inland lakes in central and S Europe. Females and juveniles, collectively known as redheads, migrate further south than males. In some winters large concentrations occur in the Netherlands. It feeds on fish and insects which it catches by diving. In its summer home it nests in holes in trees and will use nestboxes.

adult male

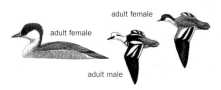

adult female

adult female

adult male

58–66 cm

Goosander

Mergus merganser

ID FACT FILE

Size: Larger than Mallard

Male: White with large green head and black back

Female: Grey body, drooping crest; contrast between reddish head, white throat and grey neck

Eclipse and Juvenile: Similar to female

Bill: Long and thin

In flight: White wing-patches (male's larger than female's)

Voice: Hoarse *kar-r-r*

Lookalikes: Red-breasted Merganser, Great Crested Grebe, Eider

The largest 'sawbill' which breeds near northern lakes and rivers. It migrates to more southerly lakes in winter and visits coastal waters less frequently than the Red-breasted Merganser. It will nest in crevices on the ground, but more often in holes in trees or nestboxes. Feeds on fish which it catches by diving. The 8–11 eggs hatch after 30 days. Young feed themselves and may join with other family groups. They fly after 60–70 days.

adult male

adult female

adult male

adult female

55–58 cm

J	F	M	A	M	J
J	A	S	O	N	D

ID FACT FILE

SIZE: Slightly larger than Buzzard

ADULTS: Dark brown above, white below with dark marks. White head with dark mask

JUVENILE: Spotted back

BILL: Hooked and dark

IN FLIGHT: Long wings, dark 'elbows', wings often bowed

VOICE: High-pitched whistles heard at nest

LOOKALIKES: Buzzard, Honey Buzzard

Osprey
Pandion haliaetus

Ospreys visit every continent except Antarctica. They are migrants, returning to Europe from Africa each spring. They usually nest in trees but will use cliffs. They need a regular supply of fish which they catch during spectacular, feet-first dives into lakes, rivers or the sea. The nest is a large structure of sticks which is re-used year after year. The 2 or 3 eggs hatch after 37 days and the young fly 53 days later.

adult

adult

adult (from below)

70–90 cm

| J | F | M | A | M | J |
| J | A | S | O | N | D |

White-tailed Eagle (Sea Eagle)

Haliaeetus albicilla

This huge eagle nests on cliffs, in trees or sometimes on the ground in N and E Europe. It was recently reintroduced to Scotland. It breeds near coasts, large rivers and lakes. Some birds migrate to central and S Europe. Feeds on fish, birds, mammals and carrion. Usually lays 2 eggs which hatch after 38 days. The young are cared for by both parents, fly after 70 days and are fed by parents for a further 35 days. Juveniles flock in winter.

ID FACT FILE

Size: The largest European eagle

Adults: Brown with pale head and white tail

Juvenile: Dark, spotted. Pale line on under-wing

Bill: Huge, hooked

In flight: Broad, rectangular, fingered wings and short, wedge-shaped tail. Shallow wing-beats and shallow, long glides

Voice: Yelping *kyick, kyick*

Lookalikes: Golden Eagle

adult

adult
(from below)

juvenile
(from below)

48–56 cm

| J | F | M | A | M | J |
| J | A | S | O | N | D |

Marsh Harrier
Circus aeruginosus

ID FACT FILE

Size: Buzzard-sized

Male: Dark body, grey head, grey wings and tail

Female: Dark. Crown and front of wings straw-coloured

Juvenile: Like an all-brown female

Bill: Hooked

In flight: Long tail, long wings often held in shallow 'V'

Voice: Usually silent

Lookalikes: Buzzard, Hen Harrier, Montagu's Harrier

The drainage of many European wetlands has reduced the population of this large bird of prey. It still breeds in some of the larger marshes and in river valleys where it hunts small mammals and birds. Northern populations migrate south in autumn and some reach Africa. The nest is a pile of reeds and sticks on marshy ground, and 3–8 eggs are laid 2–3 days apart. Young hatch after 35 days and fly 35 days later.

adult male

adult female

adult male
(from above)

adult female
(from above)

44–52 cm

| J | F | M | A | M | J |
| J | A | S | O | N | D |

Hen Harrier
Circus cyaneus

ID FACT FILE

SIZE: Smaller than Buzzard

MALE: Blue-grey with black wing-tips and white rump

FEMALE: Larger than male, brown with white rump

JUVENILE: Similar to female with more orange underparts

BILL: Hooked

IN FLIGHT: Rather narrow wings, forms shallow 'V' when soaring

VOICE: Usually silent

LOOKALIKES: Buzzard, Marsh Harrier, Montagu's Harrier

A bird of open, upland country, usually away from human disturbance. Northern populations move south or west to lowland or coastal areas for the winter and may roost communally. Often flies low, quartering the ground as it searches for small mammals and birds. Nests on the ground, laying 4–6 eggs which hatch after 30 days. Incubation starts from the first egg so ages of young are staggered. Young fly after 35 days.

adult female

adult male

adult male
(from above)

adult female
(from above)

BIRDS OF PREY

48–62 cm

| J | F | M | A | M | J |
| J | A | S | O | N | D |

Goshawk
Accipiter gentilis

ID FACT FILE

Size: Larger than Sparrowhawk; female almost Buzzard-sized

Adult: Dark brown above, white and finely barred below. Broad bands on tail, dark head, white stripe over eye

Female: Larger than male

Bill: Hooked

In flight: Like large sparrow-hawk, with shorter wings and longer tail than Buzzard

Voice: Usually silent

Lookalikes: Sparrowhawk, Buzzard, Hen Harrier

A resident in forests throughout mainland Europe, and recently has recolonised some forests in the British Isles. A secretive hunter which catches birds and mammals after a short, fast chase. The nest of sticks is built in a large tree. The 3–4 eggs hatch after 35 days, and the young are fed and tended by the female. Young males fly after 35 days, young females, being larger, fly a few days later.

adult female

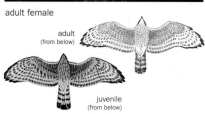

adult
(from below)

juvenile
(from below)

BIRDS OF PREY

28–38 cm

| J | F | M | A | M | J |
| J | A | S | O | N | D |

Sparrowhawk
Accipiter nisus

ID FACT FILE

Size: Kestrel-sized

Male: Blue-grey back, barred rufous under-parts

Female: Much larger, with brown back

Bill: Hooked

In flight: Broad, rounded wings, long tail. Rapid wing-beats followed by glide

Voice: Shrill *kee, kee, kee* at nest, but usually silent

Lookalikes: Goshawk, Kestrel

This small woodland hawk feeds on small birds which it pursues in flight. It never hovers like a kestrel and will hunt along woodland edges, hedges and even in gardens. Females, being larger, catch prey up to the size of pigeons. Sparrowhawks nest in trees and lay 5 eggs. Young hatch after 33 days and fly after 24–30 days, but are dependent on parents for a further 20–30 days.

adult female

adult male

adult
(from below)

juvenile
(from below)

51–57 cm

J	F	M	A	M	J
J	A	S	O	N	D

Buzzard
Buteo buteo

ID FACT FILE

SIZE: Larger than Carrion Crow

ALL BIRDS: Variable. Brown back, paler underparts; may have pale crescent on breast

BILL: Hooked

IN FLIGHT: Broad, rounded wings, short neck, shortish tail, pale patch at base of flight feathers

VOICE: Mewing call

LOOKALIKES: Honey Buzzard, Golden Eagle, Goshawk

Buzzards leave N and E Europe each autumn, elsewhere they are mainly resident. They live in cultivated country and also in uplands with wooded valleys. They feed on small mammals and other animals. Wings are held in a shallow 'V' when soaring, and the bird may hang on the wind or sometimes hover. Nests on cliffs or in trees. The 2–4 eggs hatch at 2-day intervals after 35 days. Young fly after 50 days, but are dependent on parents for a further 40–55 days.

adult

dark adult
(from below)

BIRDS OF PREY

32–35 cm

| J | F | M | A | M | J |
| J | A | S | O | N | D |

Kestrel
Falco tinnunculus

ID FACT FILE

Size: Medium-sized bird of prey

Male: Chestnut back with black spots. Blue-grey head and tail

Female: Streaky brown

Juvenile: Similar to female

Bill: Hooked, grey with yellow base

In flight: Long, pointed wings. Frequently hovers

Voice: Loud *kee-kee-kee*

Lookalikes: Merlin, Hobby, Sparrowhawk

At home in many habitats, from remote mountains to town centres. Frequently hunts beside busy roads. Found throughout Europe, but migratory in the north and east. Feeds on small mammals, birds and insects which it hunts by hovering or from a nearby perch. Nests in holes in trees, cliffs or buildings. The 3–6 eggs hatch after 27 days. Young fly after 30 days and are fed by parents for a further month.

adult male

adult female

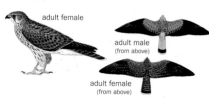

adult male
(from above)

adult female
(from above)

30–36 cm

| J | F | M | A | M | J |
| J | A | S | O | N | D |

Hobby
Falco subbuteo

This is a summer migrant to fields and open woodland of lowland Europe, where it feeds on large insects and birds, both of which it catches in flight. It is so agile that it often takes Swallows and martins and sometimes Swifts. It nests in trees, in the old nests of other species. The 3 eggs, laid in June, hatch after 28 days. Young fly after 28 days, but are dependent on parents for a further month.

ID FACT FILE

Size: Kestrel-sized

Adult: Dark blue-grey above, dark 'moustache' heavily streaked below, reddish flanks

Juvenile: Like adult but browner

Bill: Hooked

In flight: Long pointed wings give appearance of large Swift. Fast, streamlined, agile

Voice: Scolding *kew kew kew*

Lookalikes: Kestrel, Merlin, Peregrine

adult male

29–31 cm

| J | F | M | A | M | J |
| J | A | S | O | N | D |

Grey Partridge
Perdix perdix

ID FACT FILE

SIZE: Smaller than Pheasant

MALE: Plump, brown and grey, with a reddish face. Dark crescent on breast

FEMALE: Like male but with crescent reduced or absent

BILL: Pale blue-green, curved

IN FLIGHT: Rapid wing-beats followed by glide on down-curved wings

VOICE: Grating *kirr-ick*

LOOKALIKES: Red-legged Partridge, Pheasant, Quail

Open country and arable farms with hedges or other shrubs and bare areas of dry soil for dust-bathing are home for this resident game bird. Changing farming practices have caused a decline in many places. It feeds on plant material and insects, nesting on the ground in dense vegetation. The 10–20 eggs hatch after 23 days and young mainly feed themselves. They can fly by 15 days, reach their parents' weight by 100 days and stay as a family for their first winter.

adult

adults

53–89 cm
(including tail)

| J | F | M | A | M | J |
| J | A | S | O | N | D |

Pheasant

Phasianus colchicus

ID FACT FILE

SIZE: Larger than any partridge

MALE: long tail, colourful, red face, green head and neck, bronze body. Some have white ring round neck

FEMALE: Sandy-brown with darker marks

BILL: Small, pale

IN FLIGHT: Rounded wings, long pointed tail. Flies with whirring wings and a glide

VOICE: Crowing *krook-kock*

LOOKALIKES: None in area

This attractive bird has been introduced to many parts of Europe from Asia for hunting. It lives in countryside with woods and farms, and eats a wide range of food including grain, worms, spiders and green shoots. A resident, it nests on the ground among thick vegetation, laying 6–15 eggs which hatch after 23 days. Young leave the nest and feed themselves; they fly at 12 days and remain with female until about 2 months.

male breeding plumage

adult female

juvenile

adult male

adult female

23–28 cm

| J | F | M | A | M | J |
| J | A | S | O | N | D |

Water Rail
Rallus aquaticus

ID FACT FILE

SIZE: Smaller than Moorhen

ADULTS: Brown with dark streaks. Blue-grey underparts, white streaks on flanks, white under tail

BILL: Long, slightly curved, red

JUVENILE: Like adult but less well marked

IN FLIGHT: Weak with trailing legs

VOICE: Repetitive *kek* and pig-like squeals

LOOKALIKES: Moorhen, Corn-crake

The secretive and slim-bodied Water Rail walks or swims through the densely vegetated wet places where it lives. Cold winter weather sometimes forces it into more open places. It is migratory in the east, resident in the south and west, and feeds on plants and small creatures including fish. The nest is among thick vegetation on the ground. The 6–11 eggs hatch after 19 days. Young fly after 20 days. There are often 2 broods.

adult

adult

32–35 cm

| J | F | M | A | M | J |
| J | A | S | O | N | D |

Moorhen
Gallinula chloropus

ID FACT FILE

SIZE: Smaller than Coot

ADULT: Blackish, with white under tail, white stripes on flanks, long greenish-yellow legs and long unwebbed toes

BILL: Red with yellow tip

JUVENILE: Brown with pale throat and chin

IN FLIGHT: Flutters with feet dangling, but flies much more strongly when airborne

VOICE: Loud *prruck*

LOOKALIKES: Coot, Water Rail

Seldom seen far from water, Moorhens can be very secretive, but frequently they are tamer in town parks. They always seem nervous, with a stealthy walk, jerky swimming and a flickering tail. They eat a wide range of vegetable and animal food. The nest is built in or over water, and the 5–9 eggs hatch after 21 days. Young are fed by parents or by young from earlier broods. They fly after 40 days. There may be 2 or 3 broods.

adult

juvenile

adult

36–38 cm

| J | F | M | A | M | J |
| A | S | O | N | D | |

Coot
Fulica atra

ID FACT FILE

Size: Larger than Moorhen

Adults: Silky-black, with grey-green legs. Long toes partially webbed

Bill: White, with white 'shield' above bill

Juvenile: Grey with pale face and neck

In flight: Runs across water to get airborne, and often flies with trailing legs

Voice: Far-carrying *kwock*

Lookalikes: Moorhen

Quarrelsome birds of open water, Coots flock in winter when migrants from N and E Europe join residents in the south and west. They dive to feed on vegetation and small animals, but also graze on land. They are active and noisy by night as well as by day. A bulky nest is built in shallow water among vegetation. The 6–10 eggs hatch after 21 days. Young are fed by both parents for 30 days and fly at 55 days. There may be 1 or 2 broods.

adult

juvenile

adult

WADERS

40–45 cm

| J | F | M | A | M | J |
| J | A | S | O | N | D |

Oystercatcher
Haematopus ostralegus

ID FACT FILE

Size: Larger than Lapwing

All birds: Black and white, with long red legs

Winter: As summer but with white neck collar

Bill: Long, orange-red

In flight: White wing-bar on black wings

Voice: Piping *klep-klep*

Lookalikes: Avocet, Lapwing

This distinctive wader of the seashore nests inland in some places. It feeds mostly on shell-fish, especially cockles, which it extracts from the mud with its bill and prises or jabs open. It also feeds on worms. Migratory. Large flocks form in winter in W Europe. Oystercatchers nest on the ground in the open. The 3 eggs hatch after 24 days. Young are active soon after hatching, but are fed by their parents. They fly after 28 days.

adult summer

adult winter

juvenile

adult summer

WADERS

42–45 cm

J	F	M	A	M	J
J	A	S	O	N	D

Avocet
Recurvirostra avosetta

ID FACT FILE

Size: Similar to Oystercatcher

Adult: Black and white with long blue-grey legs and webbed feet

Juvenile: Dark brown rather than black on back

Bill: Upswept, black

In flight: Black and white with trailing legs

Voice: Liquid *kluit-kluit*

Lookalikes: Oystercatcher, Shelduck

Elegant, black and white wader of pools and marshes near the coast. In winter, northern birds move to sheltered estuaries in W Europe while others migrate to N Africa. The Avocet feeds on invertebrates in shallow water which it catches by sweeping its bill from side to side. It nests in the open near water and lays 2–4 eggs which take 23 days to hatch. The young feed themselves and fly after 35 days.

adult

adult

WADERS

18–20 cm

| J | F | M | A | M | J |
| J | A | S | O | N | D |

Ringed Plover
Charadrius hiaticula

ID FACT FILE

Size: Smaller than Redshank

Adult: Brown and white, with black and white head-pattern, black breast-band, orange legs

Juvenile: Scaly back, incomplete breast-band

Bill: Yellow and black

In flight: Bold white wing-bar

Voice: Piping, *tooli*

Lookalikes: Little Ringed Plover, Dotterel

The Ringed Plover feeds like other plovers – a short run followed by a quick forward tilt of the body to pick up insects or other small creatures on or near the surface. It breeds mainly near the coast, but inland in some places, and is a summer visitor to N Europe, resident in the west, and a winter visitor further south. Ground-nesting, usually among stones. Lays 3 or 4 eggs which hatch after 23 days. Young feed themselves and fly after 24 days. There are two or three broods.

adult summer

juvenile

adult summer

26–29 cm

J	F	M	A	M	J
J	A	S	O	N	D

Golden Plover
Pluvialis apricaria

ID FACT FILE

Size: Smaller than Lapwing

Summer: Brown back flecked yellow, black breast and neck, variable amount of black on face

Winter: Lacks black underparts, less yellow on back

Bill: Small, black

In flight: White underwing, dark upper wing, slight wing-bar

Voice: Lonely-sounding *too-ee*

Lookalikes: Grey Plover, Dotterel

A plover which breeds on bleak uplands and northern tundra. In winter it moves to lowland farmland in S and W Europe where hundreds, sometimes thousands, flock together. It feeds on beetles, worms and plant material. It nests on the ground, laying 4 eggs which hatch after 26 days. Young are cared for by both parents, and feed themselves. They fly after 25 days and are soon independent.

adult breeding plumage

adult winter

adult winter

adult winter

27–30 cm

J	F	M	A	M	J
J	A	S	O	N	D

Grey Plover
Pluvialis squatarola

ID FACT FILE

Size: A little larger than Golden Plover

Summer: Black belly, foreneck and face; grey spangled back

Winter: Grey spotted upperparts, pale grey underparts

Bill: Short, black

In flight: White rump, white wingbar, blackish 'armpit'

Voice: Mournful, far-carrying, *pee-oo-wee*

Lookalikes: Golden Plover

This beautiful black, white and grey wader breeds in the high Arctic and winters on muddy seashores and estuaries. In summer it feeds on insects, in winter on worms, molluscs and crustaceans. Usually solitary when feeding, but many gather in large flocks when feeding grounds are covered at high tide. Some non-breeding birds spend the summer well south of their breeding range.

adult breeding plumage

adult winter

adult winter

adult winter

28–31 cm

ID FACT FILE

Size: Smaller than Woodpigeon

All birds: Dark metallic green and white, orange under tail, thin crest, long legs

Juvenile: Shorter crest, buff tips to feathers on back

Bill: Black, stubby

In flight: Broad, rounded black and white wings. Rather floppy flight

Voice: Wheezy, drawn-out *pee-wit*

Lookalikes: None

Lapwing
Vanellus vanellus

Lapwings breed in open, flat country including farmland and coastal marshes. The aerial display is exciting and noisy as a Lapwing rises and tumbles over its territory. After nesting flocks form and travel to find suitable feeding areas free of frost. There is true migration, but severe weather triggers additional movements. Nests on the ground. The 4 eggs hatch after 26 days. Young feed themselves, fly at about 35 days and leave their parents soon after.

adult spring

juvenile

adult

20–21 cm

Sanderling
Calidris alba

ID FACT FILE

SIZE: Larger than Ringed Plover

SUMMER: Reddish-brown back, head and breast; white underparts

WINTER: Pale grey back, white underparts. Often has blackish mark on front of wings. Black legs

BILL: Black, medium length

IN FLIGHT: Dark wings, broad white wing-bar

VOICE: Quiet *kip*

LOOKALIKES: Dunlin, Knot, Little Stint

In Europe the Sanderling visits sandy beaches, running like a clockwork toy along the edge of the sea, or probing the sand around pools left by a retreating tide. It feeds on small invertebrates and often snatches food as it is washed ashore. It breeds in the high Arctic where females will sometimes lay 2 clutches of eggs, one of which is incubated by the male. Those that visit Europe are migrants from either Greenland or Siberia.

adult winter

adult summer

adult winter

WADERS

16–20 cm

| J | F | M | A | M | J |
| J | A | S | O | N | D |

Dunlin

Calidris alpina

ID FACT FILE

Size: Smaller than Redshank

All birds: Rather hunched

Summer: Rufous spotted back, grey head and neck, black belly

Winter: Grey-brown head and upperparts, pale underparts

Juvenile: Browner, scaly back and streaked breast

Bill: Black, variable length, often down-curved at tip

In flight: White wing-bar, white sides to rump

Voice: In flight a rough *treep*

Lookalikes: Little Stint, Knot, Common Sandpiper

Dunlins vary in size. The bill and leg length of northern breeding birds is noticeably shorter than that of southerly breeders. Outside the breeding season Dunlins gather in large flocks on seashores and estuaries rich in invertebrates, but also visit fringes of inland lakes. The Dunlin breeds on upland moors and on coastal grassland in the north, laying 4 eggs in a nest on the ground. The young hatch after 21 days and fly when about 20 days old.

adult winter

adult summer

juvenile

adult winter

WADERS

20–30 cm

J	F	M	A	M	J
J	A	S	O	N	D

Ruff

Philomachus pugnax

ID FACT FILE

SIZE: Male similar to Redshank, female smaller

ALL BIRDS: Long legs, small head, humped back

MALE (BREEDING): Loose ruff

MALE (NON-BREEDING): Scaly grey-brown back, buff head and breast, white underparts

FEMALE: Like small non-breeding male

BILL: Shortish and drooping

IN FLIGHT: Long-winged, loose action. White ovals on tail

VOICE: Generally silent

LOOKALIKES: Redshank, Greenshank

The extraordinary neck-ruffs are a variety of colours and used as they dance and posture at their 'leks'. Females mate with the most successful (often black-feathered) males. Ruffs are summer visitors to inland marshes, steppe and wet meadows. Most winter in Africa, while a few remain in Europe. They feed mostly on insects and their larvae. Females incubate the eggs and look after the young; 4 eggs hatch after 20 days and the young fly 25 days later.

adult male winter

adult male summer variants

adult winter

25–27 cm

J	F	M	A	M	J
J	A	S	O	N	D

Snipe
Gallinago gallinago

Snipe use their long sensitive bills to probe for worms and other invertebrates hidden in soft mud. They display over their marshland breeding territories, using their tail-feathers to produce a bleating sound called 'drumming'. Migrants leave N Europe in autumn and many winter in wet fields and marshes in the west and south. They nest on the ground and lay 4 eggs which hatch after 18 days. Young fly 19 days later.

ID FACT FILE

SIZE: Slightly smaller than Redshank

ALL BIRDS: Short legs. Brown with darker marks and buff streaks; buff stripe through centre of crown

BILL: Very long, straight

IN FLIGHT: Zigzag flight when alarmed

VOICE: Harsh *scapp* in flight, *chippa-chippa* song in spring

LOOKALIKES: Jack Snipe, Woodcock

adult

adults

33–35 cm

| J | F | M | A | M | J |
| J | A | S | O | N | D |

Woodcock
Scolopax rusticola

ID FACT FILE

SIZE: Larger than Snipe

ALL BIRDS: Plump, dark reddish-brown with delicate darker barring, black bars across crown

BILL: Very long, straight

IN FLIGHT: Display (called 'roding') is just above tree-tops, bill pointing down, owl- or bat-like, flight path often repeated

VOICE: Roding call is a series of low growls followed by sharp *twisick*

LOOKALIKES: Snipe

A secretive bird of woodlands with damp areas where it can probe for earthworms and other invertebrates. In spring males look for mates by flying slowly and low calling as they fly. Woodcocks from N Europe migrate south and west in autumn. The bird nests on the ground among low vegetation, laying 4 eggs which hatch after 22 days. Young may be able to fly from the 10th day but remain with the parent for 5–6 weeks.

adult

adults

WADERS

40–44 cm

J	F	M	A	M	J
J	A	S	O	N	D

ID FACT FILE

Size: Similar to Oystercatcher

Male (summer): Brick-red, mottled back, barred breast, white belly

Female (summer): Less colourful than male

Winter: Grey upperparts, paler underparts, pale stripe over eye

Juvenile: Similar to winter adult, but more buff, scaly back

Bill: Very long, and straight

In flight: Black tail, white rump, white wing-bars

Voice: Urgent *weeka-weeka*

Lookalikes: Bar-tailed Godwit, Spotted Redshank

Black-tailed Godwit
Limosa limosa

This large, elegant wader breeds on wet meadows, wet heaths or moorland. At other times it moves to sheltered coasts and estuaries. It feeds on insects and their larvae, worms, and seeds and other plant material. European birds migrate to S and W Europe or Africa for the winter. The nest is on the ground among short vegetation and there are 3 or 4 eggs. Young hatch after 22 days and feed themselves, but are cared for by both parents.

adult winter

adult summer

adult winter

50–60 cm

| J | F | M | A | M | J |
| J | A | S | O | N | D |

Curlew
Numenius arquata

ID FACT FILE

SIZE: The largest European wader

ALL BIRDS: Grey-brown with darker streaks and other marks, no pronounced head-pattern

MALE: Bill shorter than female

BILL: Very long, down-curved

IN FLIGHT: Pale 'V'-shaped rump

VOICE: Bubbling *cor-wee*

LOOKALIKES: Whimbrel

The bubbling song is given over upland meadows and moorlands where breeding takes place. The 'curlew' call may also be heard outside the breeding season on mudflats and sandbanks where Curlews feed on worms, crabs and other crustaceans by probing at low tide. The nest is on the ground among short vegetation; 4 eggs are incubated for 27 days. Young feed themselves and are cared for by both parents until they fly at about 32 days.

adult

adults

WADERS

30–33 cm

| J | F | M | A | M | J |
| J | A | S | O | N | D |

Greenshank
Tringa nebularia

ID FACT FILE

Size: Larger than Redshank

All birds: Upright stance, greenish-grey back. White breast and underparts are spotted in summer

Bill: Long, slightly upswept

Legs: Long, greenish

In flight: Dark wings, white 'V' on back

Voice: Ringing *tew-tew-tew*

Lookalikes: Redshank, Redshank, Ruff

A tall, elegant wader of northern bogs, often with a scattering of trees nearby. Summer migrant to N Europe, Greenshanks also visit lakes and estuaries as passage migrants usually seen singly or in very small groups. A few remain in Europe all winter, but most fly to Africa. Feeds on insects, crustaceans and fish. Nests on the ground; 4 eggs hatch after 24 days. Young feed themselves, are cared for by both parents and fly after 25 days.

adult summer

adult winter

27–29 cm

| J | F | M | A | M | J |
| J | A | S | O | N | D |

Redshank
Tringa totanus

ID FACT FILE

Size: Smaller than Lapwing

Summer: Olive-brown, heavily streaked head and breast, streaked white underparts

Winter: More uniform appearance

Bill: Slender, medium length, red base

Legs: Long, red

In flight: White trailing edges to wings, white rump

Voice: Ringing *tew*, repeated *teup*

Lookalikes: Spotted Redshank, Greenshank, Ruff

A noisy and often obvious wader which nests in wet meadows, pastures and marshes, including saltmarshes. The Redshank is usually found near the sea in winter. Its food includes shrimps, snails and worms, and it has an even-paced, jerky walk as it hunts its prey. The nest is among grasses and the 4 eggs hatch after 24 days. Young are cared for by both parents (especially the male), feed themselves and fly at 25–35 days.

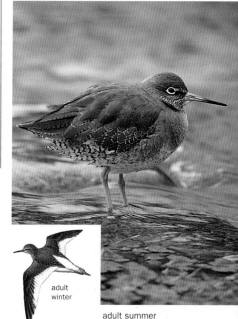

adult winter

adult summer

WADERS

22–24 cm

| J | F | M | A | M | J |
| J | A | S | O | N | D |

Turnstone
Arenaria interpres

ID FACT FILE

Size: Larger than Ringed Plover

Summer: Chestnut and black back, black and white head and breast, white belly

Winter: Blackish head and back, white underparts

Bill: Shortish, stout, black

Legs: Reddish

In flight: Striking white marks on wings, back, rump and tail

Voice: Twittering *kititit*

Lookalikes: Oystercatcher, Ringed Plover, Redshank

Picking, probing and snapping at insects as it pushes over stones or moves seaweed with its bill is the Turnstone's characteristic feeding action. It breeds mainly around the Arctic coasts and migrates to other rocky coasts, as far south as S Africa. Ground-nesting, laying 4 eggs which hatch after 22 days. Young feed themselves. Both parents care for them at first, but female may leave before they fly at 20 days.

adult winter

adult male summer

adult female summer

adult winter

34–37 cm

| J | F | M | A | M | J |
| J | A | S | O | N | D |

Black-headed Gull

Larus ridibundus

ID FACT FILE

Size: Smaller than Herring Gull

Summer: Dark brown head, pearl-grey back, black wing-tips

Winter: White head, dark mark behind eye

Juvenile: Ginger-brown marks on head and back

First winter: Dark bars on wings, black band on tail

Bill: Deep red

Legs: Deep red, feet webbed

In flight: White stripe on front edge of wings

Voice: Scolding *karrr*

Lookalikes: Little Gull, Mediterranean Gull, Common Tern

A familiar gull over much of Europe. It breeds inland or on coastal marshes and visits farmland, town parks and sheltered coasts mostly in autumn and winter when resident birds in W Europe are joined by migrants from the N and E. These gulls eat insects and worms, but also visit rubbish tips. Nests on the ground, usually in colonies. The 2 or 3 eggs hatch after 23 days. Both parents care for the young, which leave the nest after 10 days and fly 15 days later.

adult summer

adult winter

first winter

juvenile

40–42 cm

J	F	M	A	M	J
J	A	S	O	N	D

Common Gull
Larus canus

ID FACT FILE

SIZE: Smaller than Herring Gull

ADULT: White head, grey back, dark eyes. Elegant

WINTER: Heavily streaked head

FIRST WINTER: Strongly marked head, grey back, brown 'juvenile' wings

BILL: Small, yellow, no red spot

LEGS: Yellowish-green with webbed feet

IN FLIGHT: Grey back, white spots on black wing-tips

VOICE: Mewing *kee-aah*

LOOKALIKES: Herring Gull, Kittiwake

This gentle-looking gull breeds in N Europe on rocky islands, shingle bars, marshes and upland moors. It feeds on aquatic insects, worms and fish. Common Gulls migrate southwards to grasslands, agricultural land and some sea coasts. Many roost on inland reservoirs. They nest in colonies, mainly on the ground, and lay 3 eggs. Young hatch after 23 days and are cared for by both parents. They leave the nest after 5 days and fly about 30 days later.

adult summer

second winter

first winter

juvenile

52–67 cm

Lesser Black-backed Gull
Larus fuscus

ID FACT FILE

SIZE: Similar to Herring Gull but slimmer

ADULT: Elegant, white with dark grey back

SCANDINAVIAN RACE: Black back

JUVENILE: Dark brown with almost uniformly dark back

BILL: Yellow with red spot

LEGS: Yellow with webbed feet

IN FLIGHT: Rather long narrow wings. More graceful and buoyant than Herring Gull

VOICE: Gruff laughing calls

LOOKALIKES: Great Black-backed Gull, Herring Gull

Closely related to the Herring Gull, Lesser Black-backs are summer migrants to many of their northerly breeding colonies, although recently those in Britain have increasingly overwintered. Nests on islands, dunes or moors and increasingly winters in coastal areas. Eats a wide range of food, including scavenging at rubbish tips and predating other birds. The nest built of seaweed or grasses, is on the ground, and the 3 eggs hatch after 24 days. Young fly at 30–40 days.

adult summer

adult summer

second summer

juvenile

55–67 cm

| J | F | M | A | M | J |
| J | A | S | O | N | D |

Herring Gull
Larus argentatus

A familiar gull around the coasts of NW Europe. It is seldom seen far out to sea and also visits inland rubbish tips, farmland and parks. It roosts on the sea along sheltered coasts and on reservoirs. Herring Gulls eat a variety of food including carrion and offal from fishing boats. They nest on open, often sloping ground and sometimes on buildings. The 3 eggs hatch after 28 days. Young leave the nest after 3 days and fly at about 35 days.

ID FACT FILE

Size: Larger than Black-headed Gull

Adult: Pearl-grey back, fierce eye with yellow iris

Winter: Heavy streaking on back of head

Juvenile: Mottled brown, bill becoming paler and back greyer with age

Bill: Powerful, with hooked tip. Yellow with red spot

Legs: Flesh-pink with webbed feet

In flight: Broad wings, heavy-looking

Voice: Wailing, laughing, crying

Lookalikes: Common Gull, Lesser Black-backed Gull

adult summer

adult summer   second summer   first winter

64–78 cm

J F M A M J
J A S O N D

Great Black-backed Gull

Larus marinus

ID FACT FILE

Size: Large, bulky gull

Adult: Angular head, thick neck, black back with small amount of white on wing-tips

Juvenile: Dark brown chequered upperparts, paler head and tail

Bill: Long, deep and powerful. Yellow with red spot

Legs: Pink legs with webbed feet

In flight: Heavy-looking. Broad wings with large white spots at tips

Voice: Short gruff barks

Lookalikes: Lesser Black-backed Gull

The largest of the gulls eats a variety of food and also uses piracy to capture food. It frequently kills other birds, especially young seabirds. It breeds on small islands, cliff-tops and sometimes marshes or moorland. Migratory in the north, resident further south, but generally seen near the coast. The nest is a mound of vegetation. The gull lays 2 or 3 eggs which hatch after 27 days. Young are cared for by both parents and fly after about 7–8 weeks.

adults summer

adult summer

second summer

first winter

36–41 cm

| J | F | M | A | M | J |
| J | A | S | O | N | D |

Sandwich Tern
Sterna sandvicensis

ID FACT FILE

Size: Larger than Black-headed Gull

Adult (spring): White, very pale grey back, black cap, untidy crest

Adult (summer): Forehead whitens, back becomes paler

Winter: White forehead, spotted crown, untidy crest

Bill: Long, black with yellow tip

In flight: Long wings, long bill, short tail streamers. Rather stiff flight

Voice: Rasping *kirrit*

Lookalikes: Common Tern, Arctic Tern, Little Tern

This large, pale tern nests in noisy colonies on mainly sandy seashores. It plunge-dives to catch surface-living fish such as sand-eels. Some birds migrate as far as southern Africa while others winter around S European coasts. Lays 1 or 2 eggs in a shallow scrape. Young hatch after 21 days and may join a large crèche. They fly after 28 days, but may depend on their parents for the first 4 months.

adult summer

adult

juvenile

31–35 cm

Common Tern
Sterna hirundo

J	F	M	A	M	J
J	A	S	O	N	D

ID FACT FILE

Size: Smaller than Black-headed Gull

Summer: Pale grey back, large flat head, black cap

Winter: White forehead

Juvenile: Browner back, shorter wings and tail

Bill: Red with black tip

Legs: Short, red

In flight: Long pointed wings, long tail streamers, dark wedge mark on outer primaries and translucent inner primaries

Voice: Rasping *kee-aaarr*

Lookalikes: Black-headed Gull, other terns

An elegant summer visitor to beaches, fresh-water marshes and flooded sand and gravel quarries. Winters along the African coast. Catches small fish by plunge-diving. Adults display over their breeding territories. Nests on the ground usually, in colonies, and lays 1–3 eggs which hatch after 21 days. The young leave the nest after a few days and are fed by both parents. They fly after 25 days, but may be dependent on their parents for 2–3 months.

adult summer

adult summer

juvenile

22–24 cm

Black Tern
Chlidonias niger

ID FACT FILE

Size: Smaller than Black-headed Gull

Summer: Sooty-black body, slate-grey wings, white under tail

Autumn: White body, black crown extending behind eye, dark shoulder mark, paler wings

Bill: Long, black

In flight: Leisurely, often swooping down to the water or hovering

Voice: Harsh *kreert*

Lookalikes: Common Tern, Little Gull

A summer visitor to freshwater marshes where it mainly feeds on insects and their larvae. Its flight is erratic and it frequently dips to pick food delicately from the surface of the water. The birds winter mainly in the coastal region of tropical W Africa and visit larger lakes and reservoirs on migration. Lays 2–4 eggs on a mat of floating vegetation. Young hatch after 21 days, fly between 15 and 25 days and are independent soon after.

adult summer

adult autumn

juvenile

38–41 cm

J	F	M	A	M	J
J	A	S	O	N	D

Guillemot
Uria aalge

ID FACT FILE

Size: Largest member of the auk family

Summer: Dark brown or black back and head, white underparts. Some birds have white 'spectacle' mark round eye

Winter: Neck and lower face becoming white

Bill: Dagger-like, dark brown

In flight: Whirring wings, legs protruding beyond tail

Voice: Growling *aaarrr*

Lookalikes: Razorbill, Black Guillemot, Puffin

Large colonies of this penguin-like seabird breed on the dramatic sea-cliffs of NW Europe. After breeding it spends the rest of the year at sea. When feeding it chases fish underwater. The single, pear-shaped egg is laid on a cliff ledge or flat rock and hatches after 30 days. Young leave the cliff at 20 days, several weeks before being able to fly. They are fed by their parents at sea.

adult summer

adult winter

PIGEONS

32–34 cm

| J | F | M | A | M | J |
| J | A | S | O | N | D |

Stock Dove
Columba oenas

ID FACT FILE

Size: As town pigeon

All birds: Blue-grey, with purple sheen on neck, pinkish breast, slate-grey wings with short black bars

Bill: Small, pale

In flight: No white on wings but 2 short black bars. Wings held in 'V' in display

Voice: Hollow-sounding *ooo-woo*

Lookalikes: Woodpigeon, Rock Dove

Found in parkland, avenues of old trees, and copses where it nests in holes in trees or in cliff- or rock-faces. Likes water nearby, where it frequently drinks. Feeds mainly on seeds, buds and leaves. Forms flocks in winter, sometimes with Woodpigeons. A migrant in N and E Europe, resident elsewhere. Lays 2 eggs which hatch after 16 days. Young fly between 20 and 30 days. There are 2 or more broods in a year.

adult

juvenile

adult

PIGEONS

40–42cm

| J | F | M | A | M | J |
| J | A | S | O | N | D |

Woodpigeon
Columba palumbus

ID FACT FILE

SIZE: The largest pigeon

ADULT: Heavy-looking, blue-grey, with small head and pinkish breast; white marks on neck

JUVENILE: Lacking white on neck

BILL: Yellow, pink base

IN FLIGHT: White crescents on wings. 'Wing-claps' in display; glides on down-curved wings

VOICE: Gently cooing *too-COO-woo-woo*

LOOKALIKES: Stock Dove, Turtle Dove

A successful woodland species, but also familiar on farmland and, more recently, in towns and gardens. N and E European Woodpigeons are migrants and fly southwest in winter when flocks gather on agricultural land and form large roosts. Eats seeds, leaves and other plant material. Lays 1–2 eggs in a flimsy nest of twigs. Young hatch after 17 days and fly at 30 days, but may leave earlier if disturbed. There are usually 2 broods.

adult

juvenile

31–33 cm

| J | F | M | A | M | J |
| J | A | S | O | N | D |

Collared Dove
Streptopelia decaocto

ID FACT FILE

SIZE: Smaller than town pigeon

ALL BIRDS: Pinkish-grey with black half-collar

BILL: Small, dark

IN FLIGHT: Dark wing-tips, black and white pattern under tail, pale outer tail feathers. Para-chuting display

VOICE: Monotonous *co-coo-cok*

LOOKALIKES: Turtle Dove, Feral Rock Dove

This species originated in India and spread dramatically northwest across Europe during the 20th century. It is mainly resident and feeds on grain, other seeds, berries and grasses. It is common in towns, villages, parks and gardens. It lays 2 eggs in a flimsy nest of twigs in a tree or bush. Young hatch after 14 days, fly at 14 days and are independent a week later. There may be 3–6 broods a year.

adult

adults

32–34 cm

J	F	M	A	M	J
J	A	S	O	N	D

ID FACT FILE

Size: Smaller than Kestrel

Adult: Grey head and upperparts, barred under-parts, long tail, long wings which droop when perched

Juvenile: Reddish-brown, barred back, white spot on nape

Bill: Small, slightly curved

In flight: Hawk-like, fast and straight, with shallow wing-beats. Pointed wings, long tail

Voice: Male has repetitive cuckoo song, female, bubbling call

Lookalikes: Kestrel, Sparrowhawk, Nightjar

Cuckoo
Cuculus canorus

The Cuckoo's song is generally welcomed as a sign of spring. Woodland, reed-beds, moorland and farmland are some of the habitats used by Cuckoos. Dunnock, Meadow Pipit, Brambling and Redstart are all hosts for Cuckoos in Europe. Up to 25 eggs are laid by one female in different nests. The young Cuckoo hatches after 12 days, systematically ejects the other eggs and flies after 17 days, but may be dependent on its hosts for several weeks.

adult

immature

33–35 cm

| J | F | M | A | M | J |
| J | A | S | O | N | D |

Barn Owl
Tyto alba

ID FACT FILE

SIZE: Smaller than Tawny Owl

ADULT: Heart-shaped facial disc, honey-coloured back, white or buff underparts, long white legs

E EUROPE: Darker face, spotted buff underparts

BILL: Curved, almost hidden in feathers

IN FLIGHT: Silent, buoyant, wavering. Often hovers

VOICE: Many calls including loud snores from nesting birds and eerie shrieks

LOOKALIKES: Short-eared Owl

A nocturnal predator which also hunts in daylight during severe weather or when feeding young. It lives in open country with some trees and also hunts over marshes, ditches and roadside verges, feeding on small mammals, birds and insects. It nests in holes in buildings, trees or cliffs. The 4–7 eggs hatch at intervals after 30 days, so a brood contains young of various ages. They fly after 50 days.

adult

adult
(eastern race)

adult

21–23 cm

J	F	M	A	M	J
J	A	S	O	N	D

Little Owl
Athene noctua

ID FACT FILE

SIZE: Similar to Starling

ALL BIRDS: Brown spotted plumage, paler streaked underparts, fierce expression, rather long legs. Bobs when curious

BILL: Hooked, yellow-green

IN FLIGHT: Undulating, often close to the ground

VOICE: Repetitive, yapping *kiew-kiew*

LOOKALIKES: Tawny Owl

This small resident owl may be seen perched in the open during daylight. It hunts at dusk, after dark and around dawn, feeding on insects, small mammals, and worms. It lives in a variety of habitats including farmland, orchards and, in S and E Europe, it is found in hilly, arid country. It nests in holes in trees, buildings and rock faces, laying 2–5 eggs which hatch after 27 days. Young fly after 30 days and are fed by parents for a further month.

adult

adult

OWLS

37–39 cm

J	F	M	A	M	J
J	A	S	O	N	D

ID FACT FILE

SIZE: Large, round-headed

NORMAL PLUMAGE: Camouflaged, mottled brown, with dark facial disc, dark eyes, soft streaked feathers

JUVENILE: Downy and flightless on leaving nest

BILL: Small, horn-coloured, curved

IN FLIGHT: Silent with frequent glides. Large head, broad rounded wings

VOICE: Sharp *kes-wik* and quavering *poo-hooo, poo-poo-hooo*

LOOKALIKES: Long-eared Owl, Short-eared Owl

Tawny Owl
Strix aluco

This plump woodland owl rarely flies in daylight and is best known for its hooting song. It feeds on small mammals and also takes birds, insects and worms. It nests in holes in trees or old nests of other large species. The 2–5 eggs hatch after 28 days. Flightless young leave the nest at about 25 days and fly a week later. They are dependent on their parents for about 3 months.

adult

adult

OWLS

35–37cm

| J | F | M | A | M | J |
| J | A | S | O | N | D |

Long-eared Owl
Asio otus

ID FACT FILE

Size: Slightly smaller than Tawny Owl

All birds: Cat-like face can change shape. Delicately marked brown and buff plumage, tufts on head, orange eyes

Bill: Small, partly hidden, horn-coloured

In flight: Silent, ear-tufts hidden. Streaked underparts, warm closely barred wings

Voice: Low moan. Young squeak like an unoiled gate!

Lookalikes: Tawny Owl, Short-eared Owl

Tall and thin when alert, this Owl becomes fluffed out when relaxed. Tufts (which are not ears!) may be raised or lowered and are hardly visible in flight. Lives in woods, but hunts mammals and birds in open country after dark. In spring it has an aerial display in which it claps its wings together. Migrates, and may form communal winter roosts. Lays 3–5 eggs in an old nest of another species. Young hatch after 25 days, leave the nest at 21 days, fly at 30 days and become independent a month later.

adult

adults

16–17 cm

| J | F | M | A | M | J |
| J | A | S | O | N | D |

Swift
Apus apus

Swifts spend more time in flight than most birds, catching insects, drinking and even sleeping on the wing. If they do land they have great difficulty becoming airborne again. Young leaving their nests in August may spend the next 2–3 years in the air. Swifts nest in holes in buildings or rock crevices. They lay 2 or 3 eggs which hatch after 19 days. Young fly at 42 days and migrate to Africa within a few days. The birds winter in central and southern Africa.

ID FACT FILE

Size: Shorter body, longer wings than Swallow

Adult: Dark brown above and below; pale throat

Juvenile: Scaly back

Bill: Short, hooked, wide mouth

In flight: Fast, often in groups. Long, narrow, pointed wings, short forked tail

Voice: Harsh screams

Lookalikes: Swallow, Sand Martin

adult

adult

juvenile

16–17 cm

| J | F | M | A | M | J |
| J | A | S | O | N | D |

Kingfisher
Alcedo atthis

ID FACT FILE

Size: Slightly larger than House Sparrow

Adult: Bright blue-green upperparts, chestnut underparts, white throat and neck-patch

Juvenile: Duller and greener than adult

Bill: Dagger-like. Female's has red base

In flight: Fast and direct

Voice: Shrill whistle given in flight

Lookalikes: None in area

A brilliant small bird with a large head and a short tail. Found near still or slow-flowing water. Migratory in N and E Europe, mostly resident elsewhere, but some move to coasts outside the breeding season. Catches small fish by plunge-diving. The nest is in an underground chamber at the end of a tunnel which is dug by both parents. The 6 or 7 eggs hatch after 19 days and the young fly about 27 days later. There are often 2 broods.

adult male

adult female

31–33 cm

| J | F | M | A | M | J |
| J | A | S | O | N | D |

Green Woodpecker
Picus viridis

A large resident woodpecker which lives in woodland, small copses, farmland and parks. It often feeds on the ground, sometimes well away from trees. On the ground it moves with a series of hops. It eats ants and other insects. It excavates a nest-hole in the trunk of a tree and lays 5–7 eggs which hatch after 17 days. Young fly at 23 days and remain with parent for up to 7 weeks.

ID FACT FILE

Size: Similar to town pigeon

Adult: Dark green back, yellow rump, pale underparts, red crown and nape

Male: Red 'moustache'

Juvenile: Similar colours but duller, spotted and barred

Bill: Long and strong

In flight: Deeply undulating. Wings close after several strong beats

Voice: Loud laughing-yapping call, occasional feeble drumming

Lookalikes: Golden Oriole

adult female

adult

WOODPECKERS

14–15 cm

| J | F | M | A | M | J |
| J | A | S | O | N | D |

ID FACT FILE

Size: Similar to House Sparrow

All birds: Black and white barred back, white underparts

Male: Red crown

Female: White forehead

Juvenile: Less clearly marked. Some red on young male's head

Bill: Short, strong

In flight: Undulating. Barred wings

Voice: High-pitched *pee-pee-pee*

Lookalikes: Great Spotted Woodpecker

Lesser Spotted Woodpecker
Dendrocopos minor

A small resident species of European woodlands, except in the far north where it is partly migratory. Feeds on tree-trunks, in branches and among leaves, eating mainly insects. Rarely seen on the ground. Excavates its own nest-hole, often on the underside of a branch, and lays 4–6 eggs which hatch after 11 days. Young fly after 18 days. After nesting individuals may join with flocks of small birds which roam through woods and hedges.

adult male

adult female

juvenile

adult male

WOODPECKERS

22–23 cm

| J | F | M | A | M | J |
| J | A | S | O | N | D |

Great Spotted Woodpecker
Dendrocopos major

ID FACT FILE

Size: Similar to Blackbird

Adult: Black and white, red under tail, white patches on back, white cheeks separated from white throat and neck by black lines

Male: Red on back of head

Juvenile: Red crown

Bill: Dark, medium-length, strong

In flight: Bounding flight. Large white wing-patches

Voice: Sharp *kik*. Drums with beak on branch in spring

Lookalikes: Lesser Spotted Woodpecker

A resident of woodlands throughout Europe, this woodpecker eats insects, nuts and seeds, but also takes eggs and young birds in spring. It climbs trees in a series of hops, and is rarely seen on the ground. The nest is in a hole in a tree which the woodpecker chisels out itself. It lays 4–7 eggs which hatch after 10–13 days. Young fly after 20 days and stay with their parents for a further week.

adult male

adult female

juvenile

adult male

18–19 cm

| J | F | M | A | M | J |
| J | A | S | O | N | D |

Skylark
Alauda arvensis

ID FACT FILE

SIZE: Smaller than Starling

ALL BIRDS: Brown streaked back, pale underparts, streaked breast, short crest

BILL: Short, stout, horn-coloured

IN FLIGHT: Broad wings with pale hind edges, white outer tail feathers. Hovers and circles while singing

VOICE: Attractive long warbling song, usually given in flight

LOOKALIKES: Woodlark, Meadow Pipit

The Skylark's song is typical of open countryside and farmland, as the bird hangs in the air, almost too high to see. It eats insects and seeds. Northern populations migrate; in autumn flocks fly south and west to feed on arable fields. The nest is on the ground, and the 3–5 eggs hatch after 11 days. Young leave the nest at 8 days, fly after 18 days and depend on parents for another week. There are 1–3, sometimes 4 broods in S Europe.

adult

adult

SWALLOWS AND MARTINS

12 cm

J	F	M	A	M	J
J	A	S	O	N	D

Sand Martin
Riparia riparia

ID FACT FILE

Size: Smaller than Swallow

All birds: Brown above, white underparts, brown breast-band

Bill: Small, black

In flight: Rather fluttering. Brown and white with short forked tail and pointed wings

Voice: Twittering, chattering song

Lookalikes: House Martin, Swallow

A summer visitor to much of Europe, but has declined in recent years due to drought in its wintering grounds in Africa and returns in early spring to feed on insects which it catches in flight, often over water. Colonial; it nests in burrows which it digs for itself in river banks and other sandy cliffs. It lays 4–6 eggs which hatch after 14 days. Young fly after 22 days and depend on parents for a further week. There are 2 broods. Flocks roost in reedbeds on migration.

adult

adults

17–19 cm

| J | F | M | A | M | J |
| J | A | S | O | N | D |

Swallow
Hirundo rustica

ID FACT FILE

Size: Smaller than Swift

Adults: Blue-black back, pale underparts, dull red chin, long forked tail

Juvenile: Much shorter tail

Bill: Short and broad

In flight: Dark back, pale underparts, pointed wings, forked tail

Voice: Sharp *chisick* flight call, twittering song

Lookalikes: House Martin, Sand Martin, Swift

This summer migrant arrives each spring from Africa. Swallows are often seen perched on wires or swooping low over meadows, pastures and open water as they feed on flying insects. They avoid woodland and towns. The saucer-shaped mud nest is built under cover, in a barn or similar building. The 4 or 5 eggs hatch after 15 days. Young fly after 20 days and are dependent on their parents for a further week or more. There are 2–3 broods.

adult

adult   juvenile

12.5 cm

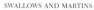

J	F	M	A	M	J
J	A	S	O	N	D

House Martin
Delichon urbica

ID FACT FILE

Size: Smaller than Swallow

All birds: Blue-black back, white underparts, white rump

Bill: Small, black

In flight: Less powerful than Swallow. Broad, pointed wings, short forked tail

Voice: *Chirrrip* call. Urgent *seep* when alarmed

Lookalikes: Swallow, Sand Martin

Most House Martins have abandoned nesting on cliffs and build their cup-shaped nests of mud under the eaves of houses in towns and villages but some cliff-nesting colonies remain. The House Martin winters in Africa. It eats insects which it catches in flight. Old nests are reused. The bird lays 3–5 eggs which hatch after 14 days, and the young fly between 22 and 32 days. There are 2 broods, and the young of the first brood may help feed the second.

adult summer

adults

14.5 cm

J	F	M	A	M	J
J	A	S	O	N	D

Meadow Pipit

Anthus pratensis

ID FACT FILE

Size: Similar to House Sparrow

All birds: Brown with darker streaks on back. Streaks on breast and flanks are of uniform size and shape. Shorter tail than Tree Pipit

Bill: Thin, pointed

In flight: Hesitant, white outer tail feathers, song given as bird rises and floats down with tail up and dangling legs

Voice: Song a series of accelerating and decelerating notes. Flight call *seep*

Lookalikes: Tree Pipit, Rock Pipit, Skylark

Meadow Pipits require no trees from which to sing. Instead they display and sing in the sky. They nest in meadowland, upland moors, lowland marshes and other open country in N Europe. In winter, northern, eastern and upland birds migrate to milder places, including farmland. Feeds on insects and plant material. Ground-nesting, laying 3–5 eggs which hatch after 13 days. Young fly at 12 days, but may leave the nest earlier. There are 2 broods.

song flight

adult

17 cm

J	F	M	A	M	J
J	A	S	O	N	D

Yellow Wagtail
Motacilla flava

ID FACT FILE

Size: Smaller than Pied Wagtail

Male: Yellow underparts, yellow, blue-grey or black head, yellow-green back. Long tail is flicked up and down

Female: Paler than male

Juvenile: Like female. Sometimes lacks yellow

Bill: Thin, pointed

In flight: Slim, long tail with white edges

Voice: Call a clear *tsweeep*

Lookalikes: Grey Wagtail, Yellowhammer

An elegant species which looks different depending on where in Europe it breeds. It winters in Africa. Head colour ranges from yellow, through blue and grey, to black. It feeds on insects and breeds in low-lying meadows and wetland margins. Nest of grass, lined with wool or fur, is built on the ground. The bird lays 4–6 eggs which hatch after 12 days. Young fly at 16 days and stay with parents for several weeks.

adult male (yellow race) summer

BLUE-HEADED RACE

adult female summer

juvenile

adult female summer

adult male summer

adult male summer   YELLOW RACE

18 cm

| J | F | M | A | M | J |
| J | A | S | O | N | D |

Pied Wagtail
Motacilla alba

ID FACT FILE

Size: Larger than House Sparrow

Adult: Black and white or grey and white. Long tail with white edges, constantly wagging up and down

Juvenile: Browner with less distinct marks, and dark patch on breast

Bill: Fine, black

In flight: Bounding, often calling

Voice: Twittering song. *Chis-ick* flight call

Lookalikes: Grey Wagtail

An adaptable species found near rivers, canals and lakes, also in towns, and often a surprising distance from water. Runs or flies to catch insects. Large numbers roost together in winter, sometimes in towns. Nests in a hole or crevice. The 5 or 6 eggs hatch after 12 days. Young fly after 13 days and parents feed them for a few more days. There are 2 broods. The race that breeds in the British Isles is darker than those found in most other parts of Europe.

adult male

adult female summer   adult male winter   juvenile   **PIED (BRITISH ISLES)**   adult female winter

adult male summer   adult male summer   adult female summer

WHITE (MAINLAND EUROPE)

9–10 cm

J	F	M	A	M	J
J	A	S	O	N	D

ID FACT FILE

Size: Smaller than Blue Tit

All birds: Tiny, dumpy; short tail often cocked above back. Brown with fine black bars, paler underparts, pale stripe over eye

Bill: Long, dark

In flight: Fast, whirring. Broad rounded wings

Voice: Fast, powerful warble ending with a trill. Call is a loud *tic-tic* or trill

Lookalikes: Dunnock, Goldcrest

Wren
Troglodytes troglodytes

One of Europe's smallest birds, which lives in many places with low cover, from small islands to mountains. Mainly resident, but migratory in the north. Severe winters may reduce numbers, but populations can recover within a few years. It eats insects, especially beetles and spiders. The male builds several domed nests, and one is chosen and lined by the female. The 5–8 eggs hatch after 16 days and young fly after 17 days. There are 2 broods.

adult

14.5 cm

| J | F | M | A | M | J |
| J | A | S | O | N | D |

Dunnock
Prunella modularis

ID FACT FILE

Size: Similar to House Sparrow

Adult: Sparrow-like body, blue-grey head and breast. Nervously flicks wings

Bill: Short, fine, blackish

In flight: Rapid, usually low. Rounded wings

Voice: Song Wren-like but slower and without trills. Loud piping call

Lookalikes: Wren, Robin, House Sparrow

In most of Europe this is a common but inconspicuous bird. It creeps like a mouse, with jerky movements. It inhabits woods and shrubberies and feeds on insects. Most territories are held by a pair of birds, but sometimes a second male assists, or the male attracts additional females. A neat nest is built in a hedge or bush. The 4–6 eggs hatch after 12 days and young fly after 11 days. There are 2 or 3 broods.

adult

juvenile

THRUSHES AND CHATS

14 cm

Robin

Erithacus rubecula

In parts of Europe Robins are shy woodland birds, elsewhere they may be quite tame and live in gardens. NE European Robins migrate and many winter around the Mediterranean. Robins feed on insects. A nest of grasses and leaves is built among tree-roots or in other sheltered positions. The 4–6 eggs hatch after 13 days, young fly 13 days later and are cared for by both parents for 16–24 days. There are 2 or 3 broods.

ID FACT FILE

SIZE: Sparrow-sized

ADULT: Plump, with short neck. Brown with red breast and face, and white belly

JUVENILE: Brown and speckled

BILL: Black and slim

IN FLIGHT: White under tail

VOICE: Fluty song, slower and sadder in autumn and winter

LOOKALIKES: Dunnock, Redstart, Bullfinch, Chaffinch

adult

juvenile

THRUSHES AND CHATS

16.5 cm

| J | F | M | A | M | J |
| J | A | S | O | N | D |

Nightingale
Luscinia megarhynchos

ID FACT FILE

Size: Larger than Robin

Adult: Rich brown upperparts, paler underparts, reddish tail, large eyes

Juvenile: Dark and light spots on back, dark spots on breast

Bill: Brown with pale base

In flight: Longer-winged and a stronger flier than Robin

Voice: Rich, varied, fluty song with deep *took-took* notes and occasional thin *seep-seep*

Lookalikes: Robin, Redstart

A skulking bird, easier to hear than to see. It sings during the day in spring, but its song is most noticeable after dark. It feeds mainly on insects and lives in woods and thickets. A summer migrant to S and central Europe. Builds a nest on or near the ground. The 4 or 5 eggs hatch after 13 days. Young fly after 11 days. In autumn it returns to tropical Africa.

adult

juvenile

14.5 cm

| J | F | M | A | M | J |
| J | A | S | O | N | D |

Black Redstart
Phoenicurus ochruros

ID FACT FILE

Size: Similar to Robin

Male (summer): Dark grey, pale patch on wings, rust-red tail

Female: Paler than male

Winter and first summer male: May be similar to female

Bill: Fine, black

In flight: Robin-like action. Red tail and rump contrast with dark body

Voice: Fast warbling song followed by metallic rattle. Repetitive *tpip* call. Hard *tic-tic* alarm

Lookalikes: Redstart

At home on remote rock-strewn mountain-sides, or in busy town centres and sometimes industrial areas where it will sing from rooftops. A summer migrant to central Europe which winters in lower-lying and coastal areas. It is resident in the south. It feeds on insects and fruit, and nests in a hole or on a ledge. The 4–6 eggs hatch after 13 days and young fly after 12 days. There are 2 broods.

adult male

adult females

14 cm

J	F	M	A	M	J
J	A	S	O	N	D

Redstart
Phoenicurus phoenicurus

ID FACT FILE

SIZE: Similar to Robin

MALE (BREEDING): Red tail with black centre, reddish under-parts, grey back, black face, white forehead

FEMALE: Brown body, paler underparts, red tail

BILL: Fine, black

IN FLIGHT: Agile. Tail looks only loosely connect-ed! Will fly from perch to snatch a flying insect; sometimes hovers

VOICE: Sweet song with mech-anical jangle at end. Call *hooveet*

LOOKALIKES: Black Redstart, Nightingale

Slimmer than a Robin, with flickering red tail. The Redstart is a summer migrant to open woodlands and parkland throughout Europe, wintering in Africa. In places it has moved into towns. It eats mainly insects which it finds on the ground or among the leaves and branches. Nest-ing in a hole in tree or, sometimes, a nestbox, it lays 5–7 eggs which hatch after 12 days. Young fly after 14 days. There are usually 2 broods.

adult male summer

adult females

THRUSHES AND CHATS

12.5 cm

Stonechat
Saxicola torquata

ID FACT FILE

SIZE: Smaller than Robin

ALL BIRDS: Short tail, plump, round head. Nervous actions

MALE (BREEDING): Blackish head, white on neck, orange-red breast

FEMALE: Lacks bold pattern of male

BILL: Small, black

IN FLIGHT: Whirring, short, rounded wings, white shoulder patch, pale rump. Sometimes hovers

VOICE: Hard *tac-tac*, like stones being struck together

LOOKALIKES: Whinchat

Stonechats require grassy areas for feeding, dense cover (often gorse) for nesting and suitable perches or song-posts. They eat insects. The Stonechat is migratory in parts of Europe, resident in others, with additional visitors in winter. It nests close to the ground. The 4–6 eggs hatch after 13 days. Young fly at 13 days and depend on both parents for a few days, then only on the male as the female prepares for another brood. There are 2 or 3 broods.

adult male summer

adult female

juvenile

adult female

THRUSHES AND CHATS

14.5–15.5 cm

J	F	M	A	M	J
J	A	S	O	N	D

Wheatear
Oenanthe oenanthe

ID FACT FILE

SIZE: Slightly larger than Robin

ALL BIRDS: White rump, short black tail. Perches upright and flicks tail

MALE (BREEDING): Black cheeks, sandy breast, blue-grey back

MALE (NON-BREEDING): Less well marked

FEMALE AND JUVENILE: Like 'washed-out' male, uniform buff-brown

BILL: Fine, black

IN FLIGHT: White rump, black tail, sometimes hovers

VOICE: Scratchy song. Harsh *chack* call

LOOKALIKES: Whinchat

A long-distance migrant from Africa to Europe; some travel even further afield, to Greenland or Siberia. Breeds in open country with bare areas such as rocky slopes, scree, tundra, cliff-tops, moors and dunes. Migrating Wheatears visit beaches and areas of short grass. The birds nest in holes, crevices or burrows. The 4–7 eggs hatch after 13 days. Young leave the nest at 10 days and fly at 15 days. There are 1 or 2 broods.

adult male breeding plumage

adult females

THRUSHES AND CHATS

24–25 cm

| J | F | M | A | M | J |
| J | A | S | O | N | D |

Blackbird
Turdus merula

ID FACT FILE

Size: Larger than Song Thrush

Male: All black, with yellow eye-ring

Female: Dark brown, pale chin, some spotting on breast

Juvenile: Reddish-brown, with spotting on upperparts and breast

Bill: Male's yellow, female's brown

In flight: Rapid, sometimes glides

Voice: Clear and fluty song which tails off at the end. Loud chucking alarm call

Lookalikes: Ring Ouzel

Some actions of a Blackbird are particularly characteristic, such as raising its tail on landing or turning over dead leaves under trees and shrubs as it searches for worms and other invertebrates. It also eats fruit, especially berries. The nest is built in a tree or bush. Between 3 and 5 eggs hatch after 13 days. Young fly after 13 days and are fed by parents for 3 weeks. There are 2 or 3 broods. Northern Blackbirds migrate south or west in autumn.

adult male

adult female

juvenile

25.5 cm

J	F	M	A	M	J
J	A	S	O	N	D

Fieldfare
Turdus pilaris

ID FACT FILE

SIZE: Slightly larger than Blackbird

ALL BIRDS: Chestnut back, dark tail, grey head and rump, bold spots on yellowish breast

BILL: Yellow, dark tip in winter

IN FLIGHT: White underwing. Burst of wing-beats followed by glides

VOICE: Chuckling song, loud *chack-chack* call often given in flight

LOOKALIKES: Mistle Thrush

A large thrush of northern woodlands of birch or conifers, but frequently moves away into open places. It has, in places, moved into towns. Feeds on invertebrates and fruits. It winters in central and S Europe and gathers into flocks. Nests in a tree, close to the trunk, laying 5 or 6 eggs which hatch after 10 days. Young fly after 12 days but depend on parents for a further 3 or 4 weeks. There are 2 broods.

adult

adult

23 cm

Song Thrush
Turdus philomelos

ID FACT FILE

SIZE: Smaller than Blackbird

ADULT: Rather short tail, brown above, pale below with many small spots

JUVENILE: Buff spots on back and head

BILL: Dark brown, yellow base

IN FLIGHT: Orange underwing, rapid and direct flight

VOICE: Song is made up of loud often fluty notes repeated several times then changed for other notes. Call is a sharp *sipp*

LOOKALIKES: Mistle Thrush, Redwing

Many species eat snails, but only the Song Thrush methodically hammers open the larger ones, often using the same stone or other hard object. It also eats other invertebrates and fruits. It lives where there are trees or bushes and open grassland. A summer migrant to N Europe, resident or winter migrant elsewhere. Nests in trees and shrubs. The 3–5 eggs hatch after 13 days and young fly 13 days later. There are 2 or 3 broods.

adult

juvenile

adult

21 cm

| J | F | M | A | M | J |
| J | A | S | O | N | D |

Redwing
Turdus iliacus

ID FACT FILE

Size: Slightly smaller than Song Thrush

All birds: Darker than Song Thrush, white stripe over eye, reddish flanks

Bill: Dark, yellowish base

In flight: Red under wing

Voice: Song is loud fluty notes followed by rapid twittering. Call is urgent *tseep*, often given in flight

Lookalikes: Song Thrush

This small thrush of northern woodlands migrates southwest in autumn. Often migrates after dark, and the high-pitched contact call can be heard as flocks fly overhead. In winter it feeds on berries or searches for worms on areas of short grass. It is vulnerable to extreme cold and will move to find milder feeding conditions in cold weather. The nest is on the ground or in a bush. The 4–6 eggs hatch after 12 days and young fly 10 days later. There are 2 broods.

adult

adult

THRUSHES AND CHATS

27 cm

| J | F | M | A | M | J |
| J | A | S | O | N | D |

Mistle Thrush
Turdus viscivorus

ID FACT FILE

Size: The largest common thrush

Adults: Larger and greyer than Song Thrush. Rather small head, pale breast with large spots, white tips to outer tail feathers

Juvenile: Spotted upperparts

Bill: Dark

In flight: Powerful, direct. White underwing

Voice: Loud, clear, with fewer notes than Blackbird. Often sings in stormy conditions. Call is a chattering rattle

Lookalikes: Fieldfare, Song Thrush

This bulky, upright thrush defends a large breeding territory in open woodland or parkland, but forms flocks in late summer. Northern populations are migratory. It eats invertebrates and fruits. In winter individuals sometimes defend a particularly good food supply such as a tree with berries. Nests early in the year, in a tree. The 3–5 eggs hatch after 12 days and young fly 12 days later. There are 2 broods.

adult

juvenile

adult

12.5 cm

| J | F | M | A | M | J |
| J | A | S | O | N | D |

Grasshopper Warbler
Locustella naevia

The Grasshopper Warbler's reeling song may be heard by day or night, but catching sight of this elusive species is difficult as it moves around in dense cover. It makes long unbroken flights to winter in Africa. Its summer home is meadowland, young plantations or fringes of wetlands. It eats insects, and builds a nest on or near the ground. The 5 or 6 eggs hatch after 12 days and young fly after 10–12 days. There are 2 broods.

ID FACT FILE

SIZE: Smaller than House Sparrow

ALL BIRDS: Faint stripe over eye, brown streaked upperparts, paler underparts, rounded tail

BILL: Fine, dark, with yellowish base

IN FLIGHT: Rounded tail sometimes obvious

VOICE: Insect-like reeling song continues for a minute or more without a break

LOOKALIKES: Cetti's Warbler, Sedge Warbler

adult

adult

WARBLERS AND CRESTS

13 cm

J	F	M	A	M	J
J	A	S	O	N	D

ID FACT FILE

Size: Smaller than House Sparrow

All birds: Streaked upper-parts, paler underparts, yellowish rump, white stripe over eye, dark crown

Bill: Blackish, with yellow base

In flight: Flits among vegetation. Song often given in short display flight

Voice: Song very varied with hard grating notes, and sparrow-like chirps

Lookalikes: Reed Warbler, Whitethroat

Sedge Warbler
Acrocephalus schoenobaenus

A summer visitor chiefly to dense vegetation growing near lakes and rivers, where it feeds on insects. In autumn it returns to central or southern Africa, making a remarkable non-stop flight across both the Mediterranean and the Sahara. The nest is usually supported by growing vegetation less than 50cm from the ground. The 5 or 6 eggs hatch after 13 days. Young fly after 13 days. There are 1 or 2 broods.

adult

adult

13 cm

J	F	M	A	M	J
J	A	S	O	N	D

Reed Warbler
Acrocephalus scirpaceus

ID FACT FILE

Size: Smaller than House Sparrow

All birds: Plain brown upperparts, paler underparts, orangeish rump, steep forehead

Bill: Long, dark, pale base

In flight: Direct over reeds

Voice: Harsh churring, repetitive, with less variety than Sedge Warbler

Lookalikes: Sedge Warbler, Garden Warbler, Cetti's Warbler

A summer visitor to stands of reeds around lakes or along rivers in many parts of Europe. It feeds on insects and spiders. Nests are woven around the stems of plants, especially common reed, usually over water. Dense, wet reedbeds may be home to large numbers of these birds. The 3–5 eggs hatch in 9–12 days. The young fly after 10 days and stay with parents for about 2 weeks. There are 1 or 2 broods. In autumn Reed Warblers return to central Africa.

adult

adult

12.5–13.5 cm

J	F	M	A	M	J
J	A	S	O	N	D

Lesser Whitethroat

Sylvia curruca

ID FACT FILE

Size: Slightly smaller than Whitethroat

All birds: Grey-brown above, pale below, white throat, grey head with dark cheeks

Bill: Small, dark

In flight: Rather compact. Direct between trees

Voice: Call is sharp *tacc*. Song is a very quiet warble followed by dry rattle

Lookalikes: Whitethroat

A rather unobtrusive warbler which sings only for a short season. It nests in hedges, bushes and small woodlands with dense cover. It feeds on insects and also berries in late summer. It winters mainly in NE Africa, making long non-stop southeastern flights with traditional stopping areas. It returns by a different route. The birds nest in small bushes. The 4–6 eggs hatch after 10 days and young fly 10 days later.

adult

adult

14 cm

| J | F | M | A | M | J |
| J | A | S | O | N | D |

Whitethroat
Sylvia communis

ID FACT FILE

SIZE: Similar to Great Tit

MALE: White throat, grey head, reddish-brown wings, pale pink or grey breast

FEMALE: Browner than male

BILL: Small, grey

IN FLIGHT: White edges to tail. Song-flight

VOICE: Call a hard *tacc*. Song an unmusical jumble of notes

LOOKALIKES: Lesser Whitethroat

A summer visitor to low, dense cover such as hedges or young plantations with patches of bramble or rose. It announces its presence with a scratchy song and a parachuting song-flight. It feeds mainly on insects. It migrates to the Sahel region of N Africa. Builds a cup-shaped nest in low bushes. The 4 or 5 eggs hatch after 11 days. Young fly after 10 days and stay with parents for 2 weeks.

adult

adult male   adult female

WARBLERS AND CRESTS

14 cm

| J | F | M | A | M | J |
| J | A | S | O | N | D |

Garden Warbler
Sylvia borin

ID FACT FILE

Size: Slightly smaller than House Sparrow

All birds: Plain brown upper-parts, paler underparts, gentle face, no obvious marks

Bill: Short, strong-looking, brownish

In flight: Direct. Square end to tail

Voice: Steady stream of melodi-ous phrases. Call *check-check*

Lookalikes: Black-cap, Reed War-bler, Chiffchaff

Woodland edge, where the undergrowth is thickest, is the summer home of this very plain, rather retiring warbler with a lovely song. It feeds on insects in summer and fruits at other times. It winters in central and southern Africa. The cup-shaped nest is built in low bushes. Its 4 or 5 eggs hatch after 11 days and young fly at 10 days, staying with parents for 2 weeks after leaving the nest.

adult

adult

13 cm

| J | F | M | A | M | J |
| J | A | S | O | N | D |

Blackcap
Sylvia atricapilla

ID FACT FILE

Size: Slightly smaller than House Sparrow

Male: Black crown, grey body, browner wings, pale underparts

Female: Grey-brown, chestnut cap

Bill: Blackish, small

In flight: Direct. Squared-off tail

Voice: Clear, rich fluty song, more varied than Garden Warbler's

Lookalikes: Garden Warbler, Whitethroat

The song of the Blackcap is a beautiful addition to mature woodlands and thickets. This warbler eats insects and fruits. In parts of W Europe the summer population migrates south to winter around the Mediterranean, and other Blackcaps from the east arrive. Sometimes these wintering birds visit gardens and bird tables. Blackcaps nest low down in dense vegetation. They lay 4–6 eggs which hatch after 11 days. Young fly at about 11 days. There are 1 or 2 broods.

adult male

adult male   adult female

WARBLERS AND CRESTS

10–11 cm

J	F	M	A	M	J
J	A	S	O	N	D

Chiffchaff
Phylloscopus collybita

ID FACT FILE

SIZE: Smaller than Blue Tit

ADULT: Brown above, pale yellowish-brown below, faint stripe over eye, dark legs. More rounded head and longer tail than Willow Warbler

JUVENILE: Tends to be more yellow than adult

BILL: Black, blunt

IN FLIGHT: Shorter, more rounded wings than Willow Warbler

VOICE: Song a repetitive *zip-zap,zip-zap*. Call a single plaintive *weet*

LOOKALIKES: Willow Warbler, Wood Warbler

A summer visitor to woods and copses over most of Europe, best known for its simple repetitive song. In autumn Chiffchaffs migrate south and west to winter around the Mediterranean and in other milder areas. Feeds on insects which it finds in trees and bushes and sometimes catches in flight. Nests in vegetation on or near the ground, laying 4–7 eggs which hatch after 13 days. Young fly at 16 days. There are 2 broods.

adult summer

adult spring

10.5–11.5 cm

J	F	M	A	M	J
J	A	S	O	N	D

Willow Warbler
Phylloscopus trochilus

ID FACT FILE

SIZE: Smaller than Blue Tit

ALL BIRDS: Like Chiffchaff, with different song, pale legs, slim rear body

ADULT: Brown-green upperparts, yellowish under-parts, pale stripe over eye. Brighter in autumn.

JUVENILE: Bright yellow underparts

BILL: Long, brown

IN FLIGHT: Longer wings and tail than Chiffchaff, sometimes hovers

VOICE: Sweet-sounding, trickling down a scale. Call is a slightly drawn-out *hooet*

LOOKALIKES: Chiffchaff, Wood Warbler

Europe's most numerous summer migrant, which winters in central and southern Africa – for some birds a journey of 12,000 km. It feeds on insects, spiders and berries. Males return first in spring and take up territory in northern birch woods, scrub, woodland edges and bushes in open country. Nests among vegetation on the ground. The 4–8 eggs hatch after 12 days. Young fly at 12 days and depend on parents for a further 2 weeks.

adult summer

adult spring

juvenile

WARBLERS AND CRESTS

9 cm

| J | F | M | A | M | J |
| J | A | S | O | N | D |

Goldcrest
Regulus regulus

ID FACT FILE

Size: Smaller than Wren

Male: Greenish with paler underparts. Dumpy with short tail. Two small wingbars, yellow and orange crown stripe, pale face and large dark eye

Juvenile: No crown stripe

Bill: Dark, small, pointed

In flight: Tiny, rapid flight on rounded wings. Short tail. Flits among branches, briefly hovers

Voice: Call a thin *see*. Song a thin rising and falling trill

Lookalikes: Firecrest, Willow Warbler

The smallest European bird. It flits restlessly from branch to branch. It breeds in conifer woods, in evergreens in parks and sometimes in deciduous woodland. After nesting, Goldcrests join flocks of small birds and may visit other habitats. Northern goldcrests migrate south to central and S Europe. The birds eat insects and spiders. The nest hangs from thin branches, and the 9–11 eggs hatch after 16 days. Young fly at 19 days. There are 2 broods.

adult

juvenile

14.5 cm

| J | F | M | A | M | J |
| J | A | S | O | N | D |

Spotted Flycatcher
Muscicapa striata

A master of flight. Darts from a perch to snatch an insect from the air and lands with a flick of its wings. A summer migrant to woodland glades and gardens with mature trees, this flycatcher winters in southern Africa. It nests in a cavity or among vegetation against a tree-trunk or wall. The 4–6 eggs hatch after 13 days. Young fly 13 days later and may depend on parents for a month. There are 1 or 2 broods.

ID FACT FILE

SIZE: Slightly smaller than House Sparrow

ADULT: Upright stance, grey-brown back, pale streaked underparts, streaked forehead and crown

JUVENILE: Buffer than adult, with spotted head, back and breast

BILL: Dark, broad base

IN FLIGHT: Swoops after flies, flits from perch to perch

VOICE: Quiet warbling song. Call is a thin squeaking *teeeze* or harder *tees-tuk-tuk*

LOOKALIKES: Dunnock, Tree Pipit

juvenile

adult

13 cm

| J | F | M | A | M | J |
| J | A | S | O | N | D |

Pied Flycatcher
Ficedula hypoleuca

ID FACT FILE

SIZE: Smaller than Spotted Flycatcher

ALL BIRDS: Round head. Flicks wings, wags tail

MALE (SPRING): Black above, white below, white patch on wing, white spot above bill

MALE (AUTUMN): Similar to female

FEMALE: Brown and white, white wing-patch

BILL: Short, black

IN FLIGHT: White wing-bar

VOICE: Sweet-sounding warble. Call a sharp *tac* or *wheeet*

LOOKALIKES: Pied Wagtail, Spotted Flycatcher

The small plump flycatcher is a summer migrant to mainly mature deciduous woods with open areas and suitable holes for nesting. It has been attracted to some woods by nestboxes. Feeds on insects, many of which it chases and catches in mid-air. Autumn migrants fly to Portugal or Spain and feed up before flying direct to W Africa. Lays 6 or 7 eggs which hatch after 13 days. Young fly after 14 days.

adult male spring

adult female

16.5 cm

J	F	M	A	M	J
J	A	S	O	N	D

Bearded Tit
Panurus biarmicus

ID FACT FILE

SIZE: Great Tit-sized with longer tail

ALL BIRDS: A similar colour to dead reeds, long-tailed

MALE: Blue-grey head, black moustache, delicate wing-pattern

FEMALE: Lacks head-pattern of male but has similar wing-pattern

BILL: Tiny, orange

IN FLIGHT: Rather weak. Whirring wings. Flies over tops of reeds. Long broad tail very obvious

VOICE: Loud *ching-ching* call

LOOKALIKES: None in area

An elusive species living in dense reed-beds where it feeds on insects, especially the larvae of moths, also spiders and seeds. It is mainly resident, but young birds roam more widely in winter. In some years, when numbers have built up, it 'erupts' and spreads to new areas. Severe winter weather can drastically reduce populations. Nests are built among reeds, and 4–8 eggs hatch after 11 days. Young fly after 12 days. There are 2–4 broods.

adult male

juvenile

adult
female summer

TITS AND ALLIES

14 cm

| J | F | M | A | M | J |
| J | A | S | O | N | D |

Long-tailed Tit
Aegithalos caudatus

ID FACT FILE

SIZE: Goldcrest-sized body, tail longer than body

ADULT: Black and white, with variable amount of reddish-brown

NORTHERN BIRDS: White head

JUVENILE: Shorter tail, browner, variable head-pattern

BILL: Short, stubby

IN FLIGHT: Weak-looking, undulating. Obvious long tail

VOICE: High-pitched *seee-seee*, low trilling *triupp*

LOOKALIKES: Pied Wagtail

Tiny body and very long tail. Mainly resident in scrub and woodland edges and visits parks and gardens. Outside the breeding season family groups roam widely and often join other families, forming flocks of over 20. At night they roost communally. The beautiful domed nest is built of moss, lichen and spiders' webs. The 8–12 eggs hatch after 15 days. Other adults may assist parents to feed their young, which fly after 14 days.

adult

juvenile

adult
northern europe

11.5 cm

J	F	M	A	M	J
J	A	S	O	N	D

Willow Tit
Parus montanus

ID FACT FILE

SIZE: Similar to Blue Tit

ALL BIRDS: Large head, pale wing-panel, dull black crown, square-ended tail. Larger black bib than Marsh Tit, and often looks fluffier

SCANDINAVIAN BIRDS: Grey and white body

SOUTHERN RACES: Brown back, off-white underparts

BILL: Small, black

IN FLIGHT: Flitting; may look weaker than Marsh Tit

VOICE: Piping *piu-piu*, buzzing *ezz-ezz-ezz*

LOOKALIKES: Marsh Tit, Coal Tit, Blackcap

A similar-looking species to Marsh Tit. Resident over much of Europe, but migratory in the far north. In places it lives in conifer woods, elsewhere in deciduous woods, especially wet woodlands with willow and alder. It eats insects and seeds. The usual nest site is a hole excavated in a rotten tree-stump. The clutch size may be influenced by the size of the cavity. The 4–11 eggs hatch after 13 days, and the young fly after 17 days.

adult

juvenile

11.5 cm

| J | F | M | A | M | J |
| J | A | S | O | N | D |

Crested Tit
Parus cristatus

ID FACT FILE

Size: Similar to Blue Tit

All birds: Rather plump with large head, black and white face and crest, brown back, pale underparts

Bill: Black

In flight: Rather strong and bouncy

Voice: Call a purring trill *zee-zee-zee*

Lookalikes: Coal Tit, Blue Tit

A forest species living mainly in pine woods but in some deciduous woods in S Europe. Searches among the highest branches for insects and spiders, but also feeds lower down and may come close to an observer. It is resident throughout the year, although juveniles sometimes rove further afield. Nests in holes in trees. The 6 or 7 eggs hatch after 13 days. Young fly after 18 days and stay with parents for about 3 weeks.

adult

juvenile

11.5 cm

| J | F | M | A | M | J |
| J | A | S | O | N | D |

Coal Tit
Parus ater

ID FACT FILE

Size: Similar to Blue Tit

Adults: Blue-grey, black head with white checks, white stripe on back of head, small double wing-bar, short tail

Juvenile: Yellowish underparts and cheeks

Bill: Quite long, pointed

In flight: Very rapid, sometimes hovers. Has short tail and wing-bars

Voice: Fast, shrill *teach-tu, teach-tu*

Lookalikes: Marsh Tit, Willow Tit, Great Tit

Conifer woods are the usual home for this small bird. Resident in S and W Europe but migratory elsewhere. Large-scale movements are sometimes triggered by shortage of food. In autumn and winter it may join flocks of other small birds and visit a variety of habitats, including gardens. Eats insects, spiders and seeds. Nests in holes, including underground. The 8–9 eggs hatch after 14 days. Young fly after 19 days. There are 1 or 2 broods.

adult

juvenile

TITS AND ALLIES

11.5 cm

J	F	M	A	M	J
J	A	S	O	N	D

Blue Tit
Parus caeruleus

ID FACT FILE

Size: Smaller than House Sparrow

Adult: Blue and green back, yellow underparts, white cheeks and blue crown

Juvenile: Like parents but less colourful, with yellow cheeks

Bill: Small and dark

In flight: Small with rounded wings

Voice: Call a shrill *tsee-tsee-tsee*

Lookalikes: Great Tit, Crested Tit

A resident woodland species. In summer it lives in lowland habitats where there are mature deciduous trees with suitable nest-holes. Will also visit other habitats, such as gardens and reed-beds, especially in winter. Feeds on insects, fruits and seeds. Breeds during April or May. A nest of moss and grasses is built in a hole in a tree, or in a nestbox. Between 6 and 16 eggs hatch after 14 days and young fly 16–22 days later. There is usually 1 brood.

adult

juvenile

14 cm

| J | F | M | A | M | J |
| J | A | S | O | N | D |

Great Tit

Parus major

ID FACT FILE

Size: Sparrow-sized

All birds: Black head, white cheeks, yellow underparts

Male: Broader black stripe on belly than female

Bill: Strong, dark grey

Juvenile: Paler version of adult

In flight: White outer tail feathers, white wing-bar

Voice: Loud repetitive *tee-cher, tee-cher*

Lookalikes: Blue Tit, Coal Tit

Lowland deciduous forest is the Great Tit's natural home, but it may breed in more open habitats with a scattering of large trees. Visits other habitats outside the nesting season. Comes to gardens for food and will sometimes nest. Feeds on insects, seeds and nuts. Nests in holes in trees and uses nestboxes. Between 3 and 18 eggs are laid in April or May. Young hatch after 14 days and leave the nest about 18 days later. There is usually 1 brood, but occasionally 2.

adult male

juvenile

14 cm

| J | F | M | A | M | J |
| J | A | S | O | N | D |

Nuthatch

Sitta europaea

ID FACT FILE

Size: Similar to Great Tit

All birds: Large head, short tail, blue-grey back, buff underparts, thick black line through eye

Bill: Long, stout, pointed

In flight: Pointed head, short square tail, broad rounded wings

Voice: Call a short loud *tuit,tuit*. Song a fast rattling *pee-pee-pee-pee*

Lookalikes: Lesser Spotted Woodpecker, Great Tit

Resident in deciduous woodland over much of Europe and, less frequently, in pine woods. On tree-trunks and branches it moves with little jumps and often descends head-first. It eats insects and seeds, including nuts which it wedges in a crevice and hammers open. Nests in holes and plasters the entrance with mud to deter other species from entering. The 6–8 eggs hatch after 14 days. Young fly after 23 days.

adult

adult male
Northern Europe

adult female
Western Europe

12.5 cm

J	F	M	A	M	J
J	A	S	O	N	D

ID FACT FILE

Size: Smaller than Nuthatch

All birds: Brown stripy back, white underparts, ragged stripe over eye, stiff pointed tail feathers

Bill: Long, down-curved

In flight: Fluttering, butterfly-like. Swoops from top of one trunk to bottom of next. Broad orange wing-bar

Voice: High-pitched *tsee, tsee* call. Song a high-pitched series of notes ending in a flourish

Lookalikes: None in area

Treecreeper
Certhia familiaris

Small brown woodland bird which creeps like a mouse up tree-trunks and larger branches, usually spiralling round as it searches for insects and spiders and some seeds in winter. Mainly resident, but some northern tree-creepers migrate. At night treecreepers roost in crevices in tree-trunks, especially ornamental redwoods. The nest is built in a crevice, often behind loose bark. The 5 or 6 eggs hatch after 14 days and the young fly after 15 days. There are 1 or 2 broods.

adult

adult

24 cm

Golden Oriole
Oriolus oriolus

ID FACT FILE

SIZE: Similar to Blackbird

MALE: Yellow, with black wings and central tail

FEMALE AND JUVENILE: Yellowish-green, dark wings, pale and streaked underparts

BILL: Long, heavy, deep pink

IN FLIGHT: Thrush-like. Closes wings and sweeps up to land on a perch

VOICE: Clear and flutey *weela-wheeloo*

LOOKALIKES: Green Woodpecker

One of Europe's most attractive summer visitors with both beautiful plumage and a melodious song. It winters in central and southern Africa and returns in spring to nest in lowland woods and other areas of deciduous trees such as poplar plantations. It feeds on insects and berries. The delicately woven nest is built in a fork of a branch high in the tree-tops. The eggs hatch after 16 days and young fly at 16 days.

adult male

adult male   adult female   juvenile

34–35 cm

J	F	M	A	M	J
J	A	S	O	N	D

Jay
Garrulus glandarius

ID FACT FILE

SIZE: Larger than town pigeon

ALL BIRDS: Pinkish-brown body, black and white wings with blue patch, black tail, white rump, small grey and white crest (sometimes raised)

OTHER RACES: Some variation in colours

BILL: Heavy, dark

IN FLIGHT: Rowing motion on broad, rounded wings. Blue wing-patch, white rump

VOICE: Wide range of calls including loud raucous screech

LOOKALIKES: None in area

A noisy but shy woodland species. Usually associated with deciduous woodland, but inhabits conifer woods in some places and also parks and large gardens. Eats invertebrates, seeds (especially acorns), eggs and young birds. Mainly resident, but migratory in the north, and occasionally food shortages trigger large movements. Nests in a fork of a tree. The 5–7 eggs hatch after 16 days. Young fly at 21 days and are fed by parents for 8 weeks.

adult

adult

CROWS

44–46 cm

Magpie
Pica pica

ID FACT FILE

Size: Larger than Woodpigeon

Adult: Iridescent black and white, with long graduated tail

Juvenile: Shorter tail, white feathers look dirty

Bill: Strong, black

In flight: Broad rounded wings, long tail

Voice: Harsh, chattering *chack-chack, chack*

Lookalikes: None in area

A familiar bird of open countryside with trees, farmland with hedges, woodland fringes, scrub and, increasingly, parks and gardens. Eats a wide range of food including insects, seeds, berries and carrion as well as small birds and eggs. A resident, but flocks together outside the breeding season. Builds a domed nest of sticks in a tree or tall bush. The 5–7 eggs hatch after 21 days. Young fly after 24 days and stay with parents for a month or more.

adult

adult

CROWS

33–34 cm

J	F	M	A	M	J
J	A	S	O	N	D

Jackdaw
Corvus monedula

ID FACT FILE

SIZE: Smaller than Rook

ALL BIRDS: Black, grey back of head, pale eye

BILL: Short, strong, black

IN FLIGHT: Acrobatic flight. Less fingered wings than other black crows. Can look pigeon-like

VOICE: sharp *jac*, ringing *key-ow*

LOOKALIKES: Chough, Rook

Small black crow with patchy distribution. Colonial, living on farmland with livestock, on cliffs (including sea-cliffs), in woodland and in villages. Often mixes with Rooks and Starlings. Mainly a resident, except in N Europe. Eats a variety of food, including invertebrates, fruits, grain and eggs. Nests in holes in trees, cliffs and buildings. The 4–6 eggs hatch after 17 days. Young fly at 32 days and become independent after 5 weeks.

adult

adult

CROWS

44–46 cm

| J | F | M | A | M | J |
| J | A | S | O | N | D |

ID FACT FILE

SIZE: Similar to Carrion Crow

ADULT: Glossy black plumage, greyish face, steep forehead, loose feathers at top of legs.

JUVENILE: Similar to adult with dark face

BILL: Long, pointed, pale base

IN FLIGHT: Fingered wing-tips, rounded tail. Narrower wings than Carrion Crow

VOICE: Harsh *karrr*

LOOKALIKES: Carrion Crow, Raven, Jackdaw

Rook
Corvus frugilegus

Rookeries of tens, sometimes hundreds, of nests are usually built in the branches of tall trees. Northern populations migrate south and west in autumn. In winter flocks of hundreds of Rooks feed together and evening roosts may attract a thousand birds or more. Rooks feed on invertebrates, grain and carrion. The 2–6 eggs hatch after 16 days. Young fly after 30 days and continue to be fed by parents for 6 weeks.

adult

juvenile

adult

45–47 cm

| J | F | M | A | M | J |
| J | A | S | O | N | D |

Carrion/Hooded Crow

Corvus corone

ID FACT FILE

SIZE: Similar to Rook

CARRION CROW: Heavy-looking, black

HOODED CROW: Grey or pinkish-brown body, black head, black tail and wings

IN FLIGHT: Powerful and slow. Fingered ends to wings

VOICE: Deep, rasping *kwarrr*

LOOKALIKES: Raven, Rook

All-black Carrion Crows live in Asia and parts of W Europe, in between there are grey and black crows called Hooded Crows. Where races meet they often interbreed. This crow eats invertebrates, grain, small animals and carrion. It is migratory in the north, sedentary in the south. Carrion crows are less gregarious than Rooks, but flocks and communal roosts do occur. The nest of sticks is built in a tree. The 3–6 eggs hatch after 18 days and young fly after 32 days.

adult hybrid

adult Carrion

adult Hooded

Carrion

Hooded

immature Carrion (right)

adult Hooded (below)

37–42 cm

| J | F | M | A | M | J |
| J | A | S | O | N | D |

Starling
Sturnus vulgaris

ID FACT FILE

Size: Smaller than Blackbird

All birds: Oily greenish-black. Short tail

Adult (winter): Pale tips to feathers give spotted appearance to back and breast

Adult (summer): Appears black, less spotted

Juvenile: Brown, paler underparts, almost white chin

Bill: Long, yellow when breeding, otherwise brown

In flight: Fast, direct, with pointed triangular wings

Voice: Scratchy whistles and warbles

Lookalikes: Blackbird

Starlings probe the ground with long, strong bills in their search for insects, but they also eat berries and other fruit. A summer visitor to N Europe, but the south and west receive migrants in autumn. Social all year, and thousands roost together in winter. Nests in holes in trees, cliffs or buildings. The 4–5 eggs hatch after 12 days. Young fly at 21 days and join flocks of other young Starlings. There are 2 broods.

adult summer

adult male winter

immature

juvenile

14–15 cm

| J | F | M | A | M | J |
| J | A | S | O | N | D |

House Sparrow
Passer domesticus

ID FACT FILE

SIZE: Slightly larger than Tree Sparrow

MALE: Brown streaked upperparts, pale underparts, grey crown, black throat, pale cheeks, grey rump

FEMALE AND JUVENILE: Duller, without prominent head markings. Straw-coloured stripe behind eye

BILL: Short, stubby

IN FLIGHT: Bounding, whirring wings

VOICE: Loud *chirrup*

LOOKALIKES: Tree Sparrow

Originally from central Asia, the House Sparrow's association with man has taken it to every continent except Antarctica. It eats grain and many other foods. A resident which flocks in autumn and winter where food is plentiful. Nests in holes, often in buildings, but sometimes builds an untidy, domed nest in a tree or bush. The 3–5 eggs hatch after 12 days. Young fly after 14 days and feed themselves 7 days later. There are 1–4 broods.

adult male spring

adult female

adult male winter

SPARROWS, BUNTINGS AND FINCHES

14.5 cm

J	F	M	A	M	J
J	A	S	O	N	D

Chaffinch
Fringilla coelebs

ID FACT FILE

Size: Sparrow-sized

Male: Pink breast and face, rest of head blue-grey. Two bold white wing-bars

Female: Shades of grey-brown, white wing-bars

Bill: Blue-grey, short and thick

In flight: White wing-bars and outer tail feathers very prominent

Voice: Loud *tink*. Song a musical rattle with flourish at end

Lookalikes: Brambling, Bullfinch

A familiar bird of woodland and other areas with trees in central and S Europe. Northern Chaffinches migrate in autumn, swelling numbers in the south and west. In autumn flocks gather with other species on farmland. The Chaffinch eats seeds and insects. It builds a neat nest of moss and lichen among branches or against a tree-trunk. The 4 or 5 eggs hatch after 12 days. Young fly at 14 days and depend on parents for 3 weeks.

adult male

adult females

14 cm

Brambling
Fringilla montifringilla

A summer migrant to northern birch forests and other woodland, where it eats mainly insects. A winter migrant to farmland and woodland in central and S Europe, feeding on seeds, especially beech mast. It flocks with other small birds or forms its own flocks which may be small or, occasionally, very large. The cup-shaped nest of moss and lichen is built against a tree-trunk or in a fork. The 5–7 eggs hatch after 11 days. Young fly at 13 days.

ID FACT FILE

SIZE: Similar to House Sparrow

MALE (SUMMER): Black head, orange breast and shoulders, white wing-bar

MALE (WINTER): Duller, with mottled head

FEMALE: Less colourful than male; orange shoulders and grey nape-patch

BILL: Stubby, pale with dark tip in winter, darker in summer

IN FLIGHT: White rump, dark tail, white wing-bar. End of tail forked

VOICE: Grating repetitive song. Wheezing flight call

LOOKALIKES: Chaffinch

adult male summer

adult male summer

adult male winter

adult female summer

adult female summer

15 cm

Greenfinch
Carduelis chloris

ID FACT FILE

Size: Similar to House Sparrow

Male: Green, green-brown wings, yellow on wings and tail

Female: Duller, streaked back, yellow on wings and tail

Juvenile: Similar to female but heavily streaked

Bill: Strong, pale

In flight: Yellow on wings and tail, green rump

Voice: Twittering flight call. Twittering song ends with nasal *dwzeeeee*

Lookalikes: Siskin, Goldfinch, Crossbill

Resident in most of Europe, but northern-populations migrate. The Greenfinch's large bill allows it to open seeds of various sizes including peanuts hung up in gardens for tits. It breeds in loose colonies in woodland, parks and large gardens. In winter it is widespread and forms flocks in open areas including arable fields. A bulky nest is built in a thick shrub. The 4–6 eggs hatch after 13 days. Young fly at 14 days. There are 2 broods.

adult male

adult female

adult female

juvenile

12 cm

SPARROWS, BUNTINGS AND FINCHES

J	F	M	A	M	J
J	A	S	O	N	D

Goldfinch
Carduelis carduelis

ID FACT FILE

Size: Smaller than Chaffinch

Adult: Buff with white belly, red, white and black face, black and yellow wings, black tail

Juvenile: Lacks red on face

Bill: Pale, narrower than most finches

In flight: Bouncing flight. Broad yellow wing-bar

Voice: Tinkling twittering call in flight. Pretty, tinkling, trilling song

Lookalikes: None in area

This beautiful small finch has a longer, more pointed bill than that of the Greenfinch, and uses it to extract seeds from food plants such as thistles. It summers in lowland Europe where there is open country with trees for nesting in. In autumn many move south towards the Mediterranean. The neat, deep nest is built towards the end of a branch. The 4–6 eggs hatch after 11 days and the young fly after 13 days. There are 1–3 broods.

adult

juvenile

adult

13.5 cm

J	F	M	A	M	J
J	A	S	O	N	D

Linnet
Carduelis cannabina

ID FACT FILE

Size: Smaller than House Sparrow

All birds: Short-tailed, brown above, paler below

Male (spring): Red forehead, grey head, red breast, plain brown back

Male (winter): Lacks red marks

Female: More streaked than winter male

Bill: Grey, small

In flight: Sparrow-like, white flashes in wings and tail

Voice: Twittering and warbling song. Twittering flight call

Lookalikes: Redpoll, Twite

Open countryside, farmland and lowland heath are home for the Linnet. It feeds almost exclusively on seeds of weeds and other plants. There is some migration and flocks form in winter, often joining with other small birds. Groups of Linnets often nest in close proximity. The nest of twigs, roots and moss is built in dense cover. The 4–6 eggs hatch after 11 days and young fly after 11 days. There are 2 or 3 broods.

adult male

adult male winter

adult female

adult female

SPARROWS, BUNTINGS AND FINCHES

12 cm

| J | F | M | A | M | J |
| J | A | S | O | N | D |

Siskin
Carduelis spinus

A small finch of conifer forests. In autumn it migrates, and flocks of Siskins are more wide-spread, feeding on birch and alder seeds and sometimes visiting gardens to take peanuts. They hang like tits to reach their food. Numbers vary considerably from year to year. The nest is small, compact and built out on a branch. The 3–5 eggs hatch after 12 days and the young fly after 13 days. There are 1 or 2 broods.

ID FACT FILE

SIZE: Smaller than Greenfinch

ALL BIRDS: Streaky yellowish-green, short forked tail, yellow in wings and tail

MALE (SPRING): Black bib and crown

MALE (WINTER): Head-pattern less distinct

FEMALE: Greyer and more streak-ed than male

BILL: Short, pointed. Male's is pale, others darker

IN FLIGHT: Light, bouncy. Yellow on wings, tail and rump

VOICE: Sweet twittering song. Call is clear *tsuu*

LOOKALIKES: Green-finch

adult male

adult female

adult female

juvenile

16.5 cm

| J | F | M | A | M | J |
| J | A | S | O | N | D |

Crossbill
Loxia curvirostra

ID FACT FILE

SIZE: Larger than Greenfinch

ALL BIRDS: Large head, short forked tail

MALE: Varies from orange-red to green, dusky wings and tail

FEMALE: Green-grey, with yellow-green rump

JUVENILE: Like female but heavily streaked

BILL: Large, heavy-looking, crossed at tip

IN FLIGHT: Powerful, bounding. Large head and short forked tail

VOICE: Greenfinch-like twittering song. *Glip-glip* call

LOOKALIKES: Greenfinch, Scottish Crossbill

Crossbills live in conifer woodlands. They move if food is short, sometimes resulting in new woods being colonised. The bill is uniquely adapted for removing seeds from cones. Nesting is linked to the cone crop and egg-laying may take place even in winter. The nest is built high in a conifer. The eggs hatch after 14 days. Young fly after 20 days and are fed by parents for 3–6 weeks.

adult male

adult female summer

adult female

juvenile

14.5–16.5 cm

J	F	M	A	M	J
J	A	S	O	N	D

Bullfinch

Pyrrhula pyrrhula

ID FACT FILE

SIZE: Larger than House Sparrow

ADULT: White rump, black tail, black wings with pale wing-bar. Grey back, black cap

MALE: Bright pink underparts

FEMALE: Pinkish-grey underparts

JUVENILE: Browner, lacks black cap

BILL: Thick, short, black

IN FLIGHT: White rump. Fluttering in bushes, strong across open ground

VOICE: Quiet warbling song. Soft, sad *pew* call

LOOKALIKES: Chaffinch

This attractive species is unpopular with gardeners and fruit-growers because it eats buds and shoots. It also eats fruits and seeds, and young Bullfinches feed on invertebrates. Northern and eastern populations migrate southwest. Bullfinches nest in woods, thickets and hedges. In winter they may visit town gardens. The nest of small twigs is built in shrubs or trees. The 3–6 eggs hatch after 12 days. Young fly after 15 days, and there are 2 or 3 broods.

adult male

adult female

juvenile

adult female

16.5 cm

J	F	M	A	M	J
J	A	S	O	N	D

Yellowhammer
Emberiza citrinella

ID FACT FILE

SIZE: Larger than House Sparrow

MALE: Bright yellow head and breast, chestnut rump

FEMALE: More variable, less yellow and more stripy

JUVENILE: Even less yellow than female

BILL: Strong, thick, bluish

IN FLIGHT: Long tail forked at tip. Strong and quite direct

VOICE: Call a sharp *zit*. Song a rattling *chitty, chitty, chitty, chee-ezz*

LOOKALIKES: Yellow Wagtail, Cirl Bunting

Resident in open countryside with bushes and scattered trees. Characteristically, it sings from the top of a shrub or from overhead wires. Yellowhammers form loose flocks in winter and northern populations migrate south. They feed mainly on seeds, but some invertebrates are eaten in summer. The nest is on or close to the ground among vegetation. The 3–5 eggs hatch after 13 days. The young are cared for by both parents and fly 11 days later. There are 2 broods.

adult male summer

adult females

15–16.5 cm

| J | F | M | A | M | J |
| J | A | S | O | N | D |

Reed Bunting
Emberiza schoeniclus

ID FACT FILE

SIZE: Slightly larger than House Sparrow

MALE (SUMMER): Black head, white collar, sparrow-like body, white outer tail feathers

MALE (WINTER): Head-pattern more like female

FEMALE: Lacks black head. Has brown cheeks and pale 'moustache'

BILL: Stubby, blackish

IN FLIGHT: Direct but a little weak. White outer tail feathers

VOICE: Call a soft *seeoo*. Song a repetitive *zinc zinc zinc zonk*

LOOKALIKES: Yellowhammer

The male often sings from a prominent perch in a reed-bed or wet ditch, but Reed Buntings also nest in drier conditions including arable fields. They eat seeds and insects. Northern populations migrate southwest in autumn. The birds form flocks in winter and often join with other small birds on farmland and other open areas. They nest among vegetation on the ground, laying 4 or 5 eggs which hatch after 13 days. Young fly after 10 days, and there are 2 broods.

adult male spring

adult females

HOP AND HEMP FAMILY, CANNABACEAE

J	F	M	A	M	J
J	A	S	O	N	D

Hop
Humulus lupulus

ID FACT FILE

HEIGHT: 3–6 m

FLOWERS: In greenish, hanging clusters, the male branched, the female cone-like with pale green bracts, each on separate plants

LEAVES: In opposite pairs, up to 10 cm long, heart-shaped, 3- to 5-lobed, coarsely toothed

FRUITS: Cone-like, hanging heads, about 3 cm long

LOOKALIKES: Hemp (*Cannabis sativa*) is an erect, often branched annual 1–3 m tall, the leaves compound, with 6–9 segments, of waste ground and tips, but occasionally cultivated.

Rough-hairy perennial, twining and trailing on hedges, shrubs, trees or wire-netting fences; cultivated in SE England and the Vale of Evesham, but often naturalised, and sometimes truly wild in damp woods. It is widespread in Britain, but rare in Scotland and Ireland, where it is introduced. The dried fruiting heads have been used in Britain since the late Middle Ages to flavour and preserve beer; also, in pillows, to aid sleep. The young shoots can be cooked as a green vegetable.

NETTLE FAMILY, URTICACEAE

J	F	M	A	M	J
J	A	S	O	N	D

Stinging Nettle
Urtica dioica

ID FACT FILE

Height:
50–150 cm,
sometimes up to
250 cm

Flowers: Tiny, in
greenish, hang-
ing, tassel-like
clusters, the
male and female
on separate
plants

Leaves: In oppo-
site pairs, up to
10 cm long,
heart- or spear-
shaped, pointed,
regularly toothed

Fruits: 1–1.5 mm
across, each
enclosing a
single seed

Lookalikes:
Annual Nettle
(*Urtica urens*) is a
smaller, often
branched annual
up to 60 cm tall,
of cultivated land.

Erect, unbranched perennial, usually covered
with long stinging hairs; it has square stems,
arising from tough, yellow roots and creeping
underground stems, and forms large patches.
Abundant on waysides and waste ground,
around buildings, in gardens, ditches, marshes
and damp woods, especially places where the
soil is enriched by animal manure or fertilizer.
The stems are an ancient source of fibre and
the shoots provide a green vegetable. Caterpillars of several butterflies feed on the leaves.

NETTLE FAMILY, URTICACEAE

Redshank or Persicaria
Persicaria maculosa

| J | F | M | A | M | J |
| J | A | S | O | N | D |

ID FACT FILE

Height:
10–80 cm

Flowers: Pale or bright pink, massed in dense, cylindrical spikes 1–3 cm long

Leaves: Narrow, spear-shaped, pointed, often with a dark blotch

Fruits: 2–3 mm long, triangular or lens-shaped, black, shiny

Lookalikes: Pale Persicaria (*Persicaria lapathifolia*) has greenish-white or dull pink flowers and flower-stalks covered with tiny, rough, yellowish hairs.

More or less hairless, erect or sometimes sprawling, branched annual. Common on cultivated land; also on bare mud and gravel beside streams, rivers and lakes. One of the most persistent weeds of farmland, sometimes colouring crops pink where it has escaped the farmer's spray, and a common plant of gardens and disturbed, waste places. Plants from waterside habitats are often smaller and little-branched. The starch-rich fruits were formerly gathered and used as a grain.

DOCK FAMILY, POLYGONACEAE

Amphibious Bistort
Persicaria amphibia

ID FACT FILE

HEIGHT:
10–100 cm

FLOWERS: Deep
pink, massed in
dense, stout,
cylindrical spikes
on long stems

LEAVES: Floating
leaves spear-
shaped, blunt,
hairless; leaves of
land plants point-
ed, minutely hairy

FRUITS: Lens-
shaped, 2–3 mm
long, brown, shiny

LOOKALIKES: On
land may be con-
fused with Red-
shank or Pale
Persicaria. Bis-
tort (*Persicaria
bistorta*), with
narrowly triangu-
lar leaves, forms
clumps in damp
grassland in N
Britain and local-
ly elsewhere.

Usually aquatic, little-branched perennial, with
floating stems and leaves, widespread in lakes,
ponds, canals and flooded ditches. Its creeping
roots spread to form large patches that can cre-
ate a conspicuous and attractive pink band
around the margins of still or slow-moving bodies of water. This species also grows on land,
sometimes as a weed of cultivation, where the
plants are erect, have narrow, rather hairy leaves
and produce fewer, scruffier heads of flowers.

DOCK FAMILY, POLYGONACEAE

J	F	M	A	M	J
J	A	S	O	N	D

Black Bindweed
Fallopia convolvulus

ID FACT FILE

HEIGHT:
10–120 cm

FLOWERS:
Inconspicuous,
greenish-white or
greenish-pink in
small clusters or
loose spikes

LEAVES: Heart- or
arrow-shaped,
pointed

FRUITS: Matt
black, triangular
nuts up to 5 mm
long, each
enclosed in
papery remains
of perianth

LOOKALIKES:
Field Bindweed
and Hedge
Bindweed are
unrelated twining
plants with white
or pink, trumpet-
shaped flowers.

Annual, prostrate or twining plant of cultivated
land. Archaeological sites reveal that it has long
been a weed of cultivation in Britain, and its
fruits were until recently a major contaminant of
agricultural seed. It represents a significant
proportion of the seed bank of cultivated land.
Plants sometimes have winged perianths, especially on rich soil, which are similar to those of
the rarer Copse Bindweed (*Fallopia dume-
torum*), of woodland margins and hedgerows.

DOCK FAMILY, POLYGONACEAE

Sorrel
Rumex acetosa

ID FACT FILE

HEIGHT:
10–120 cm

FLOWERS: Minute, reddish or greenish, in branched spikes, the male and female on separate plants

LEAVES: Basal and lower stem leaves stalked, spear- to arrow-shaped, blunt, with a pair of downward-pointing lobes at the base; upper leaves stalkless

FRUITS: 3-sided shiny brown nuts, 2–2.5 mm long, each enclosed in brown, papery remains of perianth

LOOKALIKES: Sheep's Sorrel is smaller and the basal lobes of the leaves point outwards.

Erect perennial of grassland, road-verges, woodland rides, sand-dunes and rocky ground; it can give a reddish tint to meadows in May and June. The flowers are pollinated by the wind. The whole plant tastes of acid and the leaves can be used in salads or to flavour sauces and soups – although the true Garden Sorrel is a different species. Plants from sand-dune grassland (machair) in Scotland and western Ireland are shorter, less branched and have short white hairs on the stems and leaves.

DOCK FAMILY, POLYGONACEAE

Sheep's Sorrel
Rumex acetosella

ID FACT FILE

HEIGHT: 5–30 cm, sometimes up to 50 cm

FLOWERS: Minute, reddish, in branched spikes, the male and female on separate plants

LEAVES: Narrow, spear-shaped, with a pair of outward-pointing lobes at the base, often reddish

FRUITS: 3-sided nuts, 1–1.5 mm long, each enclosed in brown, papery remains of perianth

LOOKALIKES: Sorrel is larger and the basal lobes of the leaves point downwards rather than outwards.

Erect perennial, branched above, with creeping roots, forming sometimes quite extensive patches on heathland, dry grassland, rock outcrops, wall-tops, shingle beaches and sanddunes on nutrient-poor acid soils. On sandy or peaty soils this plant can be a persistent garden and agricultural weed. The whole plant tastes of acid. The flowers are pollinated by the wind. A very variable species: particularly distinctive are plants from dry heathland, which have very narrow, strap-like leaves.

DOCK FAMILY, POLYGONACEAE

J	F	M	A	M	J
J	A	S	O	N	D

Broad-leaved Dock

Rumex obtusifolius

ID FACT FILE

HEIGHT:
50–150 cm

FLOWERS: Tiny, green or reddish, in large, loose, leafy spikes

LEAVES: Broad, oblong, heart-shaped at base, blunt

FRUITS: Tiny triangular nuts, each enclosed in brown, papery perianth, with spiny margins and a single (rarely 3) corky wart

LOOKALIKES: Other docks; Curled Dock has spear-shaped, pointed leaves with curly margins and no spines on the fruit; it rarely forms large clumps.

Robust perennial, arising from a stout root and forming conspicuous leafy clumps on cultivated land, waste places and riverbanks. It was, before modern weed-killers, a major weed of cultivation and is listed as noxious under the Weeds Act 1959. It often crosses with Curled Dock when the two grow together. The large leaves were used to wrap butter and are said to relieve nettle stings. The flowers are pollinated by the wind, but sometimes by bumblebees.

GOOSEFOOT FAMILY, CHENOPODIACEAE

J	F	M	A	M	J
J	A	S	O	N	D

Fat Hen
Chenopodium album

ID FACT FILE

HEIGHT:
20–150 cm

FLOWERS: Tiny,
green, massed in
many dense,
branched clusters

LEAVES: Oval,
diamond- or
paddle-shaped,
greyish-green,
floury, almost
untoothed or
coarsely toothed

FRUITS: Numerous,
tiny, each contain-
ing 1 black or
brown, flattened,
glossy seed

LOOKALIKES: Sea
Beet is perennial
and has dark
green, glossy
leaves. Red
Goosefoot
(*Chenopodium
rubrum*), with
jagged-toothed
leaves, grows
around manure
heaps and on
saltmarshes.

Erect, often robust and spreading branched
annual of waste ground and cultivated land,
especially on rich soil. This is the commonest
of the dozen or so goosefoots – several of them
probably introduced – in Britain and Ireland,
all plants of disturbed or open ground. Fat
Hen was once a valued substitute for spinach
and the seeds were eaten as a grain. It is a
plant that apparently has no natural habitat and
probably evolved alongside human habitation.
A very variable species.

CARNATION AND CAMPION FAMILY, CARYOPHYLLACEAE

Greater Stitchwort
Stellaria holostea

ID FACT FILE

HEIGHT: 20–60 cm

FLOWERS: In loose clusters, white, 15–25 mm across, the 5 petals divided to half-way; 10 pale yellow stamens

LEAVES: In opposite pairs, somewhat greyish-green, narrow, rough-margined, tapered from the base to a point

FRUITS: Spherical capsules 6–8 mm long, splitting into 6 segments

LOOKALIKES: Lesser Stitchwort (*Stellaria graminea*) is more slender and straggling, with smaller petals divided to the base. There are a number of similar related species.

Slender perennial, with weak, brittle, 4-angled stems, forming patches in hedgerows and woodland rides and margins. One of the most characteristic and attractive flowers of spring, brightening waysides with its masses of flowers. Although widespread, it is local in N Scotland and W Ireland and does not occur on the most acid soils. The common name refers to the slender, thread-like flower stalks, or is perhaps an allusion to its use as a folk remedy fo a stitch or sudden pain.

CARNATION AND CAMPION FAMILY, CARYOPHYLLACEAE

All through mild winters.

Chickweed
Stellaria media

ID FACT FILE

HEIGHT: 5–40 cm

FLOWERS: White, 3–10 mm across, with 5 deeply notched petals and 3–8 reddish stamens

LEAVES: In opposite pairs, oval or spear-shaped, stalked, 1-veined, pointed

FRUITS: Egg-shaped capsules, splitting into 6, drooping when ripe

LOOKALIKES: Three-nerved Sandwort (*Moehringia trinervia*) has 3–5 veins on each leaf and grows in woodland. Thyme-leaved Sandwort (*Arenaria serpyllifolia*), with tiny stalkless leaves and erect fruits, grows on bare, dry ground.

Prostrate or sprawling annual, with weak, straggling stems marked on opposite sides by a line of hairs. It forms large patches on open ground, especially cultivated land. The plant requires and tolerates high nutrient levels and survives even around manure heaps and seabird nesting cliffs. Chickweed may be a pestilential weed, but is a cheering sight as one of the first flowers of late winter and early spring, along with Shepherd's Purse, Red Dead-nettle and Groundsel.

CARNATION AND CAMPION FAMILY, CARYOPHYLLACEAE

| J | F | M | A | M | J |
| J | A | S | O | N | D |

Ragged Robin
Lychnis flos-cuculi

ID FACT FILE

HEIGHT: 25–90 cm

FLOWERS: Deep pink, rarely white, 20–25 mm across, each of the 5 petals deeply cut into 4 lobes; calyx 10-veined

LEAVES: In opposite pairs, the lower oblong, the upper narrow, spear-shaped, pointed

FRUITS: Cylindrical capsules opening by 5 short teeth, each enclosed within persistent, red-veined calyx

LOOKALIKES: Red Campion has shallowly 2-lobed petals and 10 backward-curved capsule-teeth.

Erect, often reddish, branched perennial, of marshes, damp meadows, wet woodland clearings and rides, instantly recognisable by the ragged appearance of the pink flowers. It is much less common than formerly because of the destruction of old meadows by modern agriculture, but is increasingly popular in gardens, reflecting John Gerard's comment in his 1597 *Herball*: 'These are not used either in medicine or in nourishment: but they serve for garlands or crowns, and to decke up gardens.'

CARNATION AND CAMPION FAMILY, CARYOPHYLLACEAE

All through mild winters.

Red Campion
Silene dioica

Hairy, erect, branched biennial or perennial, of woodland, shady lanes, hedgerows and coastal cliffs. It is generally common, but scarce in Ireland and local in East Anglia. Often crossing with White Campion, especially where the habitat has been disturbed. Red Campion is a native woodland plant that has expanded its range perhaps because of adaptation to more open habitats, conferred by generations of such crossing. A handsome double-flowered variant is grown in gardens.

ID FACT FILE

HEIGHT: 20–90 cm

FLOWERS: Deep pink, 15–25 mm across, male and female on separate plants, not scented, with 5 deeply notched petals; calyx 10-veined (male) or 20-veined (female)

LEAVES: In opposite pairs, broadly spear-shaped, pointed, persisting in winter

FRUITS: Cylindrical capsules, opening by 10 curved-back teeth

LOOKALIKES: Similar but taller plants with pale pink flowers are hybrids with White Campion. Ragged Robin has deeply 4-lobed petals.

WATER-LILY FAMILY, NYMPHAEACEAE

J	F	M	A	M	J
J	A	S	O	N	D

White Water-lily
Nymphaea alba

ID FACT FILE

HEIGHT: 1–2 m

FLOWERS: White, scented, 10–20 cm across, with 15–25 petals, 4 green sepals and many large, rich yellow stamens

LEAVES: Floating, circular, green, up to 30 cm across, rather leathery

FRUITS: Large spongy, warty capsules with many seeds

LOOKALIKES: Ornamental water-lilies, many of them garden hybrids.

A hairless, aquatic perennial of lakes and other still waters, arising from a massive corky rhizome and sometimes forming large patches. The leaves and flowers float at the end of long stalks. The fruits sink after flowering and ripen below the water surface. It is widespread, but commonest in W Scotland and W Ireland. It has suffered, along with other aquatics, from collection for the garden trade, and has been replaced in some areas by escaped garden plants. It is our largest wild flower.

WATER-LILY FAMILY, NYMPHAEACEAE

J	F	M	A	M	J
J	A	S	O	N	D

Yellow Water-lily
Nuphar lutea

ID FACT FILE

HEIGHT: 1–2 m

FLOWERS: Yellow, 3–8 cm across, with 5–6 conspicuous sepals, much larger than the 20 or so petals; many large, yellow stamens

LEAVES: Floating leaves circular, green, up to 40 cm across, rather leathery; underwater leaves thinner, crumpled

FRUITS: Flask-like capsules, smelling somewhat of alcohol

LOOKALIKES: Fringed Water-lily (*Nymphoides peltata*), with smaller leaves and flowers, each with 5 fringed petals, is locally common in S and C England.

A hairless, aquatic perennial of still and slow-moving waters, arising from a massive corky rhizome and sometimes forming large patches. The leaves are both submerged and floating, at the end of long, spongy stalks; the flowers emerge several centimetres above the surface of the water. The fruits ripen above the water surface. This plant is a conspicuous, characteristic and attractive feature of many streams and rivers throughout Britain and Ireland, except for much of Scotland.

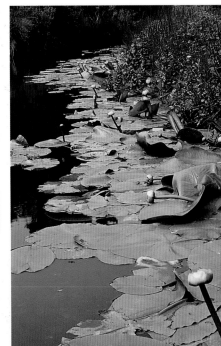

BUTTERCUP FAMILY, RANUNCULACEAE

Marsh Marigold or Kingcup
Caltha palustris

ID FACT FILE

HEIGHT:
20–50 cm

FLOWERS: Golden yellow, shiny, 2–5 cm across, with 5 petal-like sepals and many yellow stamens

LEAVES: Kidney- to heart-shaped, up to 10 cm across, dark green, the margins with neat, round teeth

FRUITS: Group of 5–15 several-seeded, beaked fruits 10–18 mm long

LOOKALIKES: The buttercups have distinct sepals and petals, and 1-seeded fruits.

Hairless perennial with stout, hollow, branched stems, forming clumps and patches in marshes, damp fields, ditches and wet woods. Widespread but less common than formerly because of the drainage of wetlands and the destruction of old meadows by modern agriculture. The plant is poisonous and avoided by grazing animals. Plants in mountainous areas are small and the stems sometimes root; they flower later. Gardeners grow orange and double-flowered variants.

BUTTERCUP FAMILY, RANUNCULACEAE

J	F	M	A	M	J
J	A	S	O	N	D

Until June in the mountains.

Wood Anemone
Anemone nemorosa

ID FACT FILE

HEIGHT: 8–25 cm

FLOWERS: Solitary, white, tinged pink or purple underneath, rarely lilac or blue, 2–4 cm across, with 5–7 (sometimes up to 12) petal-like sepals and many pale yellow stamens

LEAVES: Each with 3 much-divided and toothed lobes; stem leaves forming a protective ruff below the flower

FRUITS: 1-seeded, minutely hairy, in a round head

LOOKALIKES: Cultivated anemones, but these often have blue flowers.

Dainty, downy, erect perennial, with a creeping rhizome; locally common in open woodland, coppices and hedgerows, also meadows and mountain ledges. It is characteristic of surviving or former ancient woodland. It occurs through most of the British Isles, but is more local over much of Ireland and is absent from N Scotland. An attractive and early spring flower, it sometimes appears in great crowds, the flowers dancing in the breeze. The whole plant is poisonous.

BUTTERCUP FAMILY, RANUNCULACEAE

Meadow Buttercup
Ranunculus acris

ID FACT FILE

Height:
30–100 cm

Flowers: Bright
yellow, shiny,
15–25 mm
across, with 5
petals, in loose,
branched clus-
ters; the sepals
spreading

Leaves: Deeply
divided into 3, 5
or 7 lobes, each
deeply divided

Fruits: 1-seeded,
with hooked beak,
in a round head

Lookalikes:
Other buttercups.
Bulbous Butter-
cup has flowers
with down-curved
sepals; Creeping
Buttercup
(*Ranunculus
repens*), with
vigorous, rooting
runners, grows in
shadier places.

Erect, rather hairy perennial, a common plant
of damp grassland, road-verges, woodland
rides, marshes and mountain ledges. The commonest buttercup, sometimes colouring fields
yellow, although less so than formerly owing to
the destruction of old grassland by modern
agriculture. The whole plant is poisonous, with
acrid sap that can blister the skin. The bur-
nished appearance of buttercup flowers is
derived from light reflecting from starch grains
within the structure of the petals.

BUTTERCUP FAMILY, RANUNCULACEAE

Lesser Celandine
Ranunculus ficaria

ID FACT FILE

HEIGHT: 5–30 cm

FLOWERS: Bright yellow, shiny, 15–30 mm across, with 8–12 rather narrow petals, solitary, opening only in bright sun; and 3 sepals

LEAVES: Heart-shaped, rounded or blunt, deeply notched at base, shallowly toothed, often mottled with purplish or pale blotches

FRUITS: Tiny, 1-seeded, without a beak, in a round head; often not developing

LOOKALIKES: The closely related buttercups and crowfoots have 5 sepals and divided leaves; they mostly flower later.

Low-growing, hairless perennial, forming extensive patches in damp places in woods, hedgerows, churchyards, gardens, banks of rivers and streams and sometimes grassland. The roots develop swollen tubers, which serve as a means of reproduction and, later in the season, tuber-like bulbils often develop in the angle of the leaf-stalks and stems. One of the very first flowers to appear at winter's end and a welcome sign of coming spring. The plant dies down by the end of June.

BUTTERCUP FAMILY, RANUNCULACEAE

Celery-leaved Buttercup
Ranunculus sceleratus

ID FACT FILE

HEIGHT:
15–50 cm

FLOWERS: Pale yellow, shiny, 5–10 mm across, with 5 petals, in loose branched clusters

LEAVES: Basal leaves deeply 3-lobed, the lobes divided; stem leaves less divided; all pale green, shiny

FRUITS: Tiny, 1-seeded, hairless, in a short, cylindrical head

LOOKALIKES: Several buttercups have small, pale yellow flowers, although Celery-leaved Buttercup is the commonest.

Erect, often much-branched, mostly hairless annual, with stout, hollow stems; a common plant of muddy, trampled and open places in marshes, damp grassland and the margins of ponds, lakes and rivers. It is poisonous, especially at flowering time, with acrid sap that has a bitter, burning taste and readily blisters the mouth and skin. This plant is the main cause of buttercup poisoning amongst farm animals; it is, however, safe to feed them hay containing dried buttercups.

BUTTERCUP FAMILY, RANUNCULACEAE

J	F	M	A	M	J
J	A	S	O	N	D

Pond Water-crowfoot
Ranunculus peltatus

ID FACT FILE

HEIGHT: 10–50 cm

FLOWERS: Solitary, on stalks 50 mm or more long, white, shiny, 15–30 mm across, with 5 petals, each yellow-spotted at the base

LEAVES: Floating leaves kidney-shaped or almost round, 5-lobed (sometimes 3- or 7-lobed), glossy; submerged leaves divided into thread-like segments

FRUITS: 1-seeded, with small beak, hairy, in a round head

LOOKALIKES: Water-crowfoots differ from buttercups by their white, not yellow flowers.

Trailing, aquatic perennial that can form large populations in still, unpolluted waters of slow streams, lakes and ponds. The flowers, emerging on long, erect stalks above the surface of the water, can be an impressive sight on a sunny day. This is one of a group of ten or so water-crowfoots that occur in Britain and Ireland: some in still or slowly flowing water, with both floating and submerged leaves; some in fast-flowing water, with submerged leaves only; others on mud, without submerged leaves.

POPPY FAMILY, PAPAVERACEAE

Common, or Corn, Poppy

Papaver rhoeas

| J | F | M | A | M | J |
| J | A | S | O | N | D |

ID FACT FILE

HEIGHT: 20–80 cm

FLOWERS: Solitary, on long stalks, rich scarlet, flimsy, 5–10 cm across; many bluish-black stamens; the 2 sepals fall off

LEAVES: Compound, divided into many toothed lobes

FRUITS: Smooth, nearly spherical capsules 1–2 cm long, which release seed from a ring of pores – like a pepper pot

LOOKALIKES: Long-headed Poppy (*Papaver dubium*) has club-shaped capsules up to 25 mm long and paler-scarlet flowers. Two other, rarer, poppies have bristly fruits.

A hairy annual, with erect or ascending, bristly stems. The plant sometimes appears in great crowds, colouring cornfields and other arable land, new road-verges, allotments and waste ground red, especially on lime-rich soils in S and E England. Seed can persist in the soil for decades, germinating on exposure to light. Poppies (and other weeds) covered the disturbed chalk soil of the Somme and other battlefields after 1916, becoming a symbol of the carnage. When cut, the stems and leaves exude a white juice.

POPPY FAMILY, PAPAVERACEAE

J	F	M	A	M	J
J	A	S	O	N	D

Greater Celandine

Chelidonium majus

ID FACT FILE

HEIGHT: 20–80 cm

FLOWERS: Yellow, flimsy, 20–25 mm across, with 4 petals and 2 sepals; 2–8 in loose, domed clusters; many yellow stamens

LEAVES: Compound, with 5–9 bluntly toothed leaflets, slightly greyish-green beneath

FRUITS: Smooth, cylindrical capsules 3–5 cm long

LOOKALIKES: Welsh Poppy (*Meconopsis cambrica*) – native in rocky and shady places in W Britain and elsewhere an escape from gardens – has solitary flowers up to 4 cm across.

A tufted perennial of walls, hedge-banks and shady waste ground. Although apparently native on some rocky woodland banks, the Greater Celandine is rarely found far away from houses or ruins; it is local and certainly introduced in Scotland and Ireland. When cut, the stems and leaves exude an acrid, orange juice, which has been used to treat warts and also, more surprisingly perhaps, sore eyes. Lesser Celandine is an unrelated and quite different plant of the buttercup family.

CABBAGE AND CRESS FAMILY, CRUCIFERAE

J	F	M	A	M	J
J	A	S	O	N	D

Hedge Mustard
Sisymbrium officinale

ID FACT FILE

HEIGHT:
20–100 cm

FLOWERS: Petals
2–4 mm, pale
yellow; flowers in
narrow spikes,
lengthening con-
siderably as the
fruits develop

LEAVES: Deeply
cut into spear-
shaped lobes,
the terminal one
larger and
triangular

FRUITS: Short,
tapered pods up
to 20 mm long,
closely pressed
to the stem

LOOKALIKES: Other
members of the
family with yellow
flowers; the pale
yellow colour is
distinctive.

Roughly downy annual, or biennial, with stiff,
widely spreading branches. A common plant of
waste places and hedge-banks, also in gardens
and on cultivated land, it is widespread
throughout most of Britain and Ireland,
although scarce in N Scotland. This familiar
but not terribly attractive plant is easily recognised by the tiny clusters of pale yellow flowers
atop the long, narrow fruiting spikes. Like
most members of the cabbage and cress family,
the plant has an acrid, mustard taste.

CABBAGE AND CRESS FAMILY, CRUCIFERAE

J	F	M	A	M	J
J	A	S	O	N	D

All through mild winters.

ID FACT FILE

HEIGHT: 5–60 cm

FLOWERS: 2–3 mm across, in long, slender spikes; petals white, longer than the green or pink calyx

LEAVES: Oblong or spear-shaped, deeply cut into pointed lobes, the end one larger, triangular; upper leaves few, arrow-shaped, clasping the stem

FRUITS: Small, flattened, triangular or heart-shaped pods, each like a little purse

LOOKALIKES: Field Penny-cress has more showy flowers and much larger, circular fruits.

Shepherd's Purse
Capsella bursa-pastoris

An erect, downy, branched annual, with a neat rosette of basal leaves. Abundant on cultivated and waste ground and perhaps the commonest and most familiar of all weeds. It is one of the first flowers of late winter and early spring, along with Chickweed, Red Dead-nettle and Groundsel. In winter the stems and pods are attacked by white fungus, which covers and distorts them. A very variable species, especially in the size of the flowers and shape of the pods and leaves.

CABBAGE AND CRESS FAMILY, CRUCIFERAE

Garlic Mustard or Jack-by-the-Hedge
Alliaria petiolata

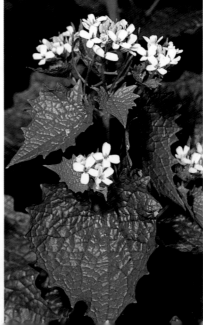

ID FACT FILE

HEIGHT:
30–120 cm

FLOWERS: Petals
2–3 mm, white;
flowers in domed
clusters, elongating as the fruits
develop

LEAVES: Long-
stalked, oval or
triangular, with
heart-shaped
base, deeply
toothed

FRUITS: Slightly
4-angled,
cylindrical pods
2–7 cm long, on
short, stiff,
spreading stalks

LOOKALIKES: The
heart-shaped
leaves and garlic
smell distinguish
this plant from
other members
of the family with
white flowers.

An almost hairless, robust biennial, or short-lived perennial, smelling strongly of garlic when bruised. Common in hedgerows, woodland margins, shady banks, churchyards and gardens. Its rather stiff, upright habit and white flowers make the plant a feature of waysides in May. This is one of the food plants of Orange Tip and Green-veined White butterflies. Garlic Mustard and a few species of penny-cress are remarkable for their garlic smell, which is rare outside the onion family.

CABBAGE AND CRESS FAMILY, CRUCIFERAE

Lady's Smock or Cuckooflower

Cardamine pratensis

A dainty, erect perennial of marshy ground, damp grassland and woodland rides, also on mountain ledges and even damp lawns. It has long been welcomed as a symbol both pure and sensual of spring, flowering as it does, in John Gerard's evocative description in his 1597 *Herball*: 'for the most part in April and May, when the Cuckoo begins to sing her pleasant note'. A particularly attractive, double-flowered variant, prized by gardeners, can sometimes be found in the wild.

ID FACT FILE

HEIGHT:
20–40 cm

FLOWERS: 1–2 cm across, lilac or almost white, in loose clusters

LEAVES: Neatly compound, the leaflets slightly toothed, the terminal one larger; stem leaves smaller, with narrower lobes

FRUITS: Cylindrical pods, 2–4 cm long, exploding to disperse the seeds

LOOKALIKES: Sea Rocket (*Cakile maritima*), also with lilac or whitish flowers, is an annual plant of seashores that has fleshy leaves and short, very stout fruits.

MIGNONETTE FAMILY, RESEDACEAE

J	F	M	A	M	J
J	A	S	O	N	D

Wild Mignonette

Reseda lutea

ID FACT FILE

HEIGHT:
30–80 cm

FLOWERS: Pale yellow in 5–20 cm spikes; 6 petals, 3–4 mm long, and 12–20 stamens

LEAVES: Deeply divided and lobed, pale green

FRUITS: Cylindrical capsules with black, shiny seeds

LOOKALIKES: Weld has strap-shaped leaves, 4 petals and 20–30 stamens.

A sprawling, often rather untidy, branched annual, biennial or perennial of disturbed ground, roadsides and waste places, usually on lime-rich soil, especially chalk. It is less frequent in the west and is rare in Scotland; in Ireland it occurs only near the east coast. The English name derives from its similarity to the scented Mignonette (*Reseda odorata*) of old-fashioned cottage gardens. The fruits, and those of Weld, are unusual because they are not fully closed, even before they are ripe.

STONECROP AND HOUSELEEK FAMILY, CRASSULACEAE

J	F	M	A	M	J
J	A	S	O	N	D

Biting Stonecrop
Sedum acre

ID FACT FILE

HEIGHT:
10–80 cm

FLOWERS: Starry, bright yellow, with 5 petals, up to 10 mm across, in terminal clusters

LEAVES: Very fleshy, narrowly egg-shaped; sharp, peppery taste

FRUITS: Fused clusters of 5 enclosed in the enlarged, persistent flower

LOOKALIKES: Other stonecrops, most of them escaped from gardens.

A low-growing, fleshy perennial, with erect shoots and flowering-stems, forming extensive patches on sand-dunes, shingle beaches, walls, railway ballast and waysides. In midsummer this plant provides spectacular splashes of yellow on disused or neglected railway sidings, concrete tracks, former airfields and car-parks. The leaves often have a red tinge, caused by lack of nutrients in the shallow soils. An alternative, if obscure, English name is 'Welcome-home-husband-however-drunk-you-be'!

Grass-of-Parnassus
Parnassia palustris

ID FACT FILE

HEIGHT:
10–50 cm

FLOWERS: Solitary, white, veined with green, 15–30 mm across, scented, with 5 petals; 5 stamens, also 5 fringed structures that attract insects

LEAVES: Mostly in basal rosette, long-stalked; heart-shaped; 1 single, clasping stem leaf

FRUITS: Egg-shaped capsules 12–20 mm long, splitting into 4 segments

LOOKALIKES: Unlikely to be confused with any other plant.

An elegant, rather sticky-hairy, erect perennial, of well-drained grassland on lime-rich soil. It used to be more common before modern farming destroyed old meadows, but can still be found here and there on dry roadside banks and especially in country churchyards. It occurs mainly in E and S England; it is scarce in Scotland and very rare indeed in Ireland. Small tuber-like bulbils at the base of the leaves allow the plant to reproduce vegetatively as well as by seed.

ROSE FAMILY, ROSACEAE

J	F	M	A	M	J
J	A	S	O	N	D

Meadowsweet
Filipendula ulmaria

ID FACT FILE

HEIGHT:
50–120 cm, sometimes up to 200 cm

FLOWERS: Creamy-white, 4–8 mm across, scented, in wide, dense, branched, frothy clusters

LEAVES: Compound, with up to 5 pairs of large leaflets, smaller leaflets between, and a 3- to 5-lobed terminal leaflet, all white-hairy beneath

FRUITS: Few-seeded, in clusters of 6–10

LOOKALIKES: Drop-wort (*Filipendula vulgaris*), of dry grassland on limestone, has leaves with 8–20 pairs of leaflets, and fewer flowers 8–18 mm across.

One of our most beautiful wild flowers: an erect, hairless perennial with several flowering stems, of marshes, wet heaths, and damp places in sand-dunes. It is local in distribution, absent from much of S England, Wales and SW Ireland; it has become extinct in the English Midlands as a result of the drainage and destruction of wetlands. This is the only European species of a N-hemisphere group of some 15 species. Plants from sand-dunes are shorter and have slightly larger flowers.

ROSE FAMILY, ROSACEAE

| J | F | M | A | M | J |
| J | A | S | O | N | D |

Dog Rose
Rosa canina

ID FACT FILE

HEIGHT: 1–3 m

FLOWERS: Varying shades of pink, 2–6 cm across, with 5 petals and many stamens

LEAVES: Compound with 3–5 oval, toothed, green leaflets

FRUITS: Fleshy, flask-shaped or almost globular, scarlet or pale orange, topped by dry remains of flower

LOOKALIKES: Field Rose (*Rosa arvensis*) is up to 1 m tall and has white flowers; Burnet Rose (*Rosa pimpinellifolia*), forming clumps of densely prickly stems, has smaller white flowers.

An arched shrub, the stems armed with ferocious, curved prickles. One of the best-loved of all wild flowers, brightening hedges with great splashes of pink in June. There are some dozen or so closely related species of wild rose in Britain and Ireland, all loosely termed dog roses, together with hybrids; several garden roses also escape and cross with native species. Dog Rose was formerly used as a stock for grafting garden roses, but other species are now used. The fruits are rich in vitamin C.

ROSE FAMILY, ROSACEAE

| J | F | M | A | M | J |
| J | A | S | O | N | D |

Bramble or Blackberry
Rubus fruticosus

ID FACT FILE

HEIGHT: 1–4 m, but mostly trailing

FLOWERS: White, pink or purplish-pink, 2–3 cm across, with 5 petals and many stamens

LEAVES: With 3 oval, coarsely toothed, green leaflets, paler beneath

FRUITS: Fleshy, globular heads, 1–2 cm across, of black, 1-seeded drupes

LOOKALIKES: Wild Raspberry (*Rubus idaeus*) has erect stems, leaflets in pairs and red, softer fruits.

An arched, prickly shrub, forming a tangle of stems near the ground and climbing into bushes and hedges. Common and widespread, it can be a serious weed of woods and waste ground, although it provides cover and food for many animals. The fruits are an ever-popular source of free food and the roots yield an orange dye; the leaves have been used to treat wounds and appear in herbal teas. Botanists divide Blackberry up into hundreds of similar 'micro-species', many of them very local in distribution.

ROSE FAMILY, ROSACEAE

| J | F | M | A | M | J |
| J | A | S | O | N | D |

Agrimony
Agrimonia eupatoria

ID FACT FILE

HEIGHT:
30–150 cm

FLOWERS: 6–8 mm across, with 5 yellow petals and 10–20 stamens, massed in a long spike

LEAVES: Basal leaves in a rosette; leaves compound, with 6–8 pairs of large and 2–3 pairs of smaller leaflets, all greyish-hairy beneath

FRUITS: In shape of inverted cone (like an old-fashioned top), grooved, half enclosed by the cup-like calyx, covered with hooked bristles

LOOKALIKES: The tall spikes of yellow flowers and bristly fruits are distinctive.

Erect, softly hairy perennial of road verges, hedgerows, scrub, woodland margins and tall grassland; conspicuous in late summer. It is widespread, but is scarce over much of Scotland, especially the north. The hooked spines of the persistent calyx that surrounds the ripe fruit readily attach themselves to fur or clothing and fruits are thus dispersed far from the parent plant. The plant yields a yellow dye and has long been used medicinally as an antiseptic and general tonic.

ROSE FAMILY, ROSACEAE

Wood Avens
Geum urbanum

An erect, branched, leafy perennial of open woodland, woodland margins and rides, hedgerows and shady gardens; widespread, but local in N Scotland. The long hooks on the fruits attach themselves to fur or clothing and the fruits are thus dispersed far from the parent plant. The flowers are pollinated by flies. This species often crosses with Water Avens in W and N Britain, with hybrids and backcrosses displaying a mixture of the features of both parents.

ID FACT FILE

HEIGHT:
20–60 cm

FLOWERS: Shallow-ly cup-shaped, yellow, 10–15 mm across, the calyx green; many stamens

LEAVES: Basal leaves in a rosette; leaves compound, with 1–5 pairs of unequal leaflets

FRUITS: Round cluster of hairy, 1-seeded fruits, each with a persistent, hooked stigma

LOOKALIKES: Water Avens has larger, nodding, bell-shaped, pink flowers and grows in damper places.

ROSE FAMILY, ROSACEAE

Silverweed
Potentilla anserina

ID FACT FILE

HEIGHT: 10–30 cm

FLOWERS: Golden yellow, 15–20 mm across, with 5 petals and many stamens, solitary on long stems

LEAVES: Leaves all in a basal rosette, compound, with 7–25 coarsely toothed, oblong or oval leaflets, silvery-hairy beneath

FRUITS: Head of small, round achenes

LOOKALIKES: Creeping Cinquefoil (*Potentilla reptans*) is a more slender, far-creeping plant, with green leaves and leaflets grouped in fives.

Silvery-hairy, prostrate perennial with far-creeping, red, rooting stems, forming extensive patches. Abundant in damp, grassy places, waste ground and seashores near the point reached by the highest tides. The fleshy roots were formerly extensively eaten, and were a staple during times of famine, especially in N and W Britain and in Ireland; it may even have been truly cultivated in highland regions. The plant has been used for its healing properties.

ROSE FAMILY, ROSACEAE

Tormentil
Potentilla erecta

A dainty perennial with slender, sprawling, ascending or weakly erect stems, common in bogs, damp grassland, birch woods, moors and heaths, especially in hilly districts and in the mountains. Its presence always indicates that the soil is acid, i.e. very low in lime. The rather woody roots yield a red dye and were formerly used in tanning. The plant sometimes forms intermediate hybrids with Creeping Cinquefoil (see LOOKALIKES).

ID FACT FILE

HEIGHT: 10–40 cm

FLOWERS: Yellow, 8–18 mm across, with 4 (rarely 5) petals and sepals and many stamens; grouped in loose, leafy clusters

LEAVES: Leaves mostly in a basal rosette, which often withers by flowering time, with 3–5 spear-shaped, sharply toothed leaflets

FRUITS: Round head of minute, hairless 1-seeded fruits

LOOKALIKES: Creeping Cinque-foil (*Potentilla reptans*) has creeping runners, leaves with 5–7 leaflets, and 4-petalled flowers 15–35 mm across.

CLOVER AND PEA FAMILY, LEGUMINOSAE

Broom
Cytisus scoparius

| J | F | M | A | M | J |
| J | A | S | O | N | D |

ID FACT FILE

Height: 1–3 m

Flowers: Golden yellow, 15–25 mm long, in short, dense clusters

Leaves: On young shoots only, each with 3 spear-shaped or narrowly oval, silky-hairy leaflets

Fruits: Flattened pods, 25–40 mm long, hairy on the margins

Lookalikes: Gorse is spiny, with shorter pods. Spanish Broom (*Spartium junceum*), escaping from gardens mostly on to railway embankments, has larger flowers in long, loose clusters.

An erect shrub, the stems 5-angled and apparently leafless. A locally common plant of open woods, heaths, cliffs, shingle beaches, dry banks and disused railway tracks. The stems were formerly important for fodder and bedding and were cut and sold in bundles for brooms. Broom and Gorse were part of the rural economy of heathland, a habitat that is notably characteristic of parts of Britain. A prostrate variant of Broom is sometimes found on cliffs in Cornwall, Wales and Ireland and on coastal shingle in Kent.

CLOVER AND PEA FAMILY, LEGUMINOSAE

J	F	M	A	M	J
J	A	S	O	N	D

Tufted Vetch
Vicia cracca

ID FACT FILE

HEIGHT:
50–200cm

FLOWERS: Bluish-purple, 8–15 mm long, in dense, 1-sided clusters of 10–40 atop a long stalk

LEAVES: Compound, with 6–15 pairs of oblong or narrowly oval leaflets and a long, branched terminal tendril

FRUITS: Flattened, brown, hairless pods 10–22 mm long

LOOKALIKES: Other vetches, but distinguished by the larger clusters of bluish-purple flowers.

Rather hairy perennial, scrambling by means of tendrils, of bushy and grassy places, especially hedgerows, scrub and woodland margins, and old meadows. This striking flower of summer is perhaps the most familiar of the several native vetches. It is rarely seen in meadows today because of their destruction by modern agriculture. Some plants are very hairy. Tufted Vetch belongs to a group of five closely related European species that can be very difficult to tell apart.

CLOVER AND PEA FAMILY, LEGUMINOSAE

J	F	M	A	M	J
J	A	S	O	N	D

Common Vetch
Vicia sativa

ID FACT FILE

HEIGHT:
20–120 cm

FLOWERS: Bright
pink, purplish-
crimson or pur-
ple, 10–15 mm
long, stalkless,
solitary or in
pairs

LEAVES: Com-
pound, with
3–8 pairs of
oblong or oval,
very narrow to
almost heart-
shaped leaflets
and a branched
terminal tendril

FRUITS: Black or
yellowish-brown,
hairless or downy
pods 25–70 mm
long

LOOKALIKES: Other
vetches, but dis-
tinguished by the
combination of
leaf-tendrils and
solitary or paired
pinkish flowers.

A hairy, climbing or trailing perennial, often
rather neat in habit by comparison with other
vetches, growing in grassy places, borders of
fields, waysides and hedge-banks, in many
places as a relic of former cultivation for
fodder. A very variable species which has been
divided by botanists into several subspecies.
The variation partly reflects a long history as a
crop, with much seed being brought in from
abroad. Plants with bright pink flowers and
narrow leaflets are probably native.

CLOVER AND PEA FAMILY, LEGUMINOSAE

J	F	M	A	M	J
J	A	S	O	N	D

Meadow, or Yellow, Vetchling
Lathyrus pratensis

ID FACT FILE

Height:
30–120 cm

Flowers: Yellow,
12–20 mm long,
in clusters of
5–12 atop a
long, erect stalk

Leaves: Paired,
spear-shaped,
entire leaflets up
to 5 cm long,
with a pair of
large, leaflet-like
structures at the
base and a
branched
terminal tendril

Fruits: Black,
hairless pods
2–4 cm long

Lookalikes:
Bird's-foot Trefoil
has no tendrils,
leaflets in fives
and smaller,
often orange-tint-
ed, flowers.

Climbing or sprawling perennial, arising from
a slender rhizome, with angled, square, slightly
winged, rather weak, leafy stems. It is common
and conspicuous in grassy places, marshes,
hedge-banks, scrub and woodland margins;
indeed, it is one of the most familiar and
attractive of all our grassland wild flowers. It
is a principal food plant for the caterpillar of
the Wood White butterfly. An infusion of this
plant has been used as a remedy for coughs
and bronchitis.

CLOVER AND PEA FAMILY, LEGUMINOSAE

Tall, or Golden, Melilot
Melilotus altissimus

ID FACT FILE

HEIGHT:
50–150 cm

FLOWERS: Yellow,
5–7 mm long, in
loose, 1-sided,
many-flowered
clusters; the
petals are all the
same length

LEAVES: Leaflets
3, wedge-shaped
or oval, toothed

FRUITS: Black,
short-hairy pods
3–5 mm long

LOOKALIKES:
Common Melilot
(*Melilotus
officinalis*) is
often taller, with
the upper 3
petals longer
than the lower
pair, and the fruit
brown and hair-
less; White
Melilot (*Melilotus
alba*) has white
flowers.

An erect, branched, often quite robust annual,
biennial or short-lived perennial that is widespread on waste ground, field borders and
agricultural land. It tends to prefer heavier
soils. It may be native, but has been spread by
cultivation as a fodder crop; it has also been
used medicinally and may have been first introduced here by 16th-century herbalists. It is rare
in Scotland, and in Ireland it occurs only near
the coasts of Co. Dublin and Co. Wicklow.

CLOVER AND PEA FAMILY, LEGUMINOSAE

J	F	M	A	M	J
J	A	S	O	N	D

Lesser Trefoil or Suckling Clover

Trifolium dubium

ID FACT FILE

Height:
10–30 cm, sometimes up to 50 cm

Flowers: Tiny, yellow, 10–20 in spherical heads 7–9 mm across, on a long stalk

Leaves: Leaflets 3, oval, toothed, the middle one stalked

Fruits: Tiny, inconspicuous, 1-seeded pods enclosed within persistent corolla

Lookalikes: Black Medick is a more robust, somewhat hairy plant with denser heads and conspicuous black fruits; the petals do not persist in fruit.

Often rather hairy, sometimes even slightly sticky-hairy, prostrate or weakly erect annual, or short-lived perennial, of dry grassland, sunny banks, walls, waysides, coastal cliffs and sand-dunes. It is common throughout, but less so in Scotland. Many plants of waste places and fields are more robust and more erect; they probably derive from former cultivation as a fodder crop or from the use of commercial, non-native wild flower seed. This plant is often confused with three native yellow-flowered clovers.

J	F	M	A	M	J
J	A	S	O	N	D

Red Clover
Trifolium pratense

ID FACT FILE

HEIGHT:
20–60 cm,
sometimes up to
100 cm

FLOWERS: Pale or
dark pink, or red-
dish-purple, rarely
white, 12–15 mm
long, in dense,
solitary or paired,
almost spherical
heads 2–4 cm
long, with 2
leaves immedi-
ately below

LEAVES: Leaflets
3, oval or almost
circular, hairy
beneath, often
marked with a
whitish crescent

FRUITS: Inconspic-
uous, 1-seeded
pods enclosed
within persistent
corolla; the fruit-
ing head is a
brown ball of
dead flowers

LOOKALIKES: The
commonest red-
flowered clover.

A rather hairy, tufted, sprawling, ascending or
erect, short-lived perennial of grasslands, waysides and waste ground throughout Britain and
Ireland. A very variable species. Many plants
derive from introduced fodder crops or wild
flower seed: these are robust, tall and erect,
and have large, often pale pink heads of flowers, whereas the true native plant has a more
prostrate or sprawling habit and smaller heads
of usually richly red or reddish-purple flowers.
The flowers are much visited by bumblebees.

CLOVER AND PEA FAMILY, LEGUMINOSAE

| J | F | M | A | M | J |
| J | A | S | O | N | D |

White, or Dutch, Clover
Trifolium repens

ID FACT FILE

HEIGHT:
20–50 cm

FLOWERS: Scented, white, often tinged pink, 8–13 mm long, in dense, solitary, long-stalked, spherical heads 1–3 cm across

LEAVES: Long-stalked; leaflets 3, oval or almost circular, each frequently marked with a whitish V or dark markings

FRUITS: Inconspicuous, 1- to 4-seeded pods, enclosed within persistent corolla; the fruiting head is a brown ball of dead flowers

LOOKALIKES: The common white-flowered clover of grassland in Britain and Ireland.

A creeping perennial, with rooting, prostrate stems, forming large patches in lawns, grasslands, waysides and waste ground throughout Britain and Ireland. A variable species, with most plants probably now derived from cultivated plants. White Clover is of considerable economic value as a fodder crop and source of nectar for bees; bacteria in nodules on the roots convert atmospheric nitrogen to usable plant nutrients. A special variety with purple flowers occurs in the Isles of Scilly.

CLOVER AND PEA FAMILY, LEGUMINOSAE

Bird's-foot Trefoil
Lotus corniculatus

ID FACT FILE

HEIGHT: 5–50 cm

FLOWERS: Yellow, often streaked or tinted orange or red, 10–18 mm long, 3–8 in clusters atop a long stalk

LEAVES: Each with 5 spear-shaped to almost circular leaflets

FRUITS: Cylindrical pods 10–15 mm long, arranged like a bird's foot

LOOKALIKES: Marsh Bird's-foot Trefoil (*Lotus uliginosus*) is taller and hairier, with calyx-teeth in bud spreading at right-angles. Meadow Vetchling has tendrils and larger, entirely yellow flowers.

A prostrate, sprawling or erect perennial, forming patches; widespread in dry grassland and on sunny banks, cliffs, rocks and sand-dunes, even lawns, throughout Britain and Ireland. It grows on well-drained soils, particularly those derived from sand or limestone. On road-verges and elsewhere plants are often more robust and erect (Fodder Bird's-foot Trefoil), and are a non-native variety derived from commercial wild flower seed. It is an important food plant for caterpillars of Blue and other butterflies.

J	F	M	A	M	J
J	A	S	O	N	D

Wood Sorrel
Oxalis acetosella

ID FACT FILE

HEIGHT:
10–20 cm

FLOWERS: Solitary
on almost leaf-
less stems, bell-
shaped, nodding,
8–15 mm long,
white with lilac
veins or some-
times all lilac

LEAVES: Long-
stalked, delicate,
pale green, often
purplish beneath;
leaflets 3, heart-
shaped, notched

FRUITS: Egg-
shaped, 5-angled
capsules,
exploding when
ripe to disperse
the seeds

LOOKALIKES: Pink
Sorrel (*Oxalis
articulata*), with
clusters of bright
pink flowers, is a
garden escape
on roadsides and
waste ground.

Delicate, slender, hairless perennial, with
creeping rhizome, clothed in swollen, scale-
like remains of leaf-bases; widespread in moist,
shady woods and on mountain-ledges. The
leaves have a sharp acid taste and were formerly used as a flavouring, like those of Sorrel.
This plant has some claim to be the *Seamróge*
or Shamrock by which St Patrick illustrated the
concept of the Holy Trinity to the then pagan
Irish; however that is more likely to have been
Lesser Trefoil.

GERANIUM FAMILY, GERANIACEAE

Dove's-foot Cranesbill

Geranium molle

J	F	M	A	M	J
J	A	S	O	N	D

ID FACT FILE

HEIGHT: 5–40 cm

FLOWERS: Dish-shaped, purplish-pink or rarely white, 5–10 mm across, in very loose terminal clusters; 5 deeply notched petals

LEAVES: Deeply divided into 3–5 3-lobed segments, softly hairy, slightly greyish-green

FRUITS: Slightly wrinkled, hairless, beak-like, with 5 segments that curl upwards explosively when ripe to disperse the seed

LOOKALIKES: The commonest of several annual cranesbills that occur in Britain and Ireland.

A hairy, branched, weak-stemmed annual of disturbed and open ground, especially waysides, waste ground and coastal sand-dunes, and even dry, unmanicured lawns. This is the commonest of a group of small, rather similar annual or short-lived perennial cranesbills, all plants of ground opened up by human activity. They are characterised by round, deeply lobed or dissected leaves, small, pink or purplish flowers and long, slender, beak-like fruits.

GERANIUM FAMILY, GERANIACEAE

Herb Robert

Geranium robertianum

ID FACT FILE

HEIGHT:
20–50 cm

FLOWERS: Dish-shaped, purplish-pink, 15–30 mm across, in very loose terminal clusters; 5 petals; stamens usually orange

LEAVES: Rather fern-like, cut almost to the base into 3–5 deeply lobed segments

FRUITS: Slightly wrinkled, beak-like, with 5 segments that curl upwards explosively when ripe to disperse the seed

LOOKALIKES: Several plants in the carrot family have similar fern-like leaves, but very different flowers.

Rather dainty, aromatic, reddish, hairy, annual, biennial or short-lived perennial with straggling, brittle, fleshy stems. It is a common plant of shady places, cliffs, rocks, walls and coastal shingle beaches; and a weed of shady gardens and old glasshouses. Plants from W Britain and Ireland sometimes have white flowers. A small-flowered subspecies of Herb Robert (known as Little Robin), with yellow stamens and more wrinkled fruits, is a choice rarity of S and W England and Co. Cork.

GERANIUM FAMILY, GERANIACEAE

J	F	M	A	M	J
J	A	S	O	N	D

Common Stork's-bill
Erodium cicutarium

ID FACT FILE

HEIGHT: 5–60 cm

FLOWERS: Dish-shaped, purplish-pink, lilac or sometimes white, 8–20 mm across, 2–12 in loose heads; 5 deeply notched petals

LEAVES: Almost fern-like, divided twice into deeply lobed segments, softly hairy or sticky-hairy

FRUITS: Slightly wrinkled, with a beak up to 6 cm long and 5 segments that twist explosively into spirals when ripe

LOOKALIKES: The cranesbills lack the spiral twisting of the fruit-segments; apart from Herb Robert, they do not have fern-like leaves.

Hairy annual, biennial or short-lived perennial, of open sandy and stony ground and thin grass-land, especially by the sea; a very common plant on sand-dunes, where it is mostly prostrate and can be sticky-hairy. It is mainly coastal in Ireland. The explosive release of the seeds serves to disperse them efficiently; each has a spiral, corkscrew-like segment of the fruit wall attached, which screws it into loose soil. A variable species, notably in the colour and size of the petals.

SPURGE FAMILY, EUPHORBIACEAE

J	F	M	A	M	J
J	A	S	O	N	D

Sun Spurge
Euphorbia helioscopa

ID FACT FILE

HEIGHT:
10–50 cm

FLOWERS: Yellowish, with complex structure characteristic of the spurges: a cluster of tiny male and female flowers within an envelope or involucre

LEAVES: Oval or spoon-shaped, toothed towards the tip

FRUITS: Smooth capsules 2–3.5 mm long

LOOKALIKES:
Petty Spurge (*Euphorbia peplus*), a branched, leafy annual with untoothed leaves, is more a weed of gardens.

An erect, mostly unbranched, yellowish-green, hairless annual of disturbed and cultivated ground, especially arable land. This plant grows best on rich soils and accumulates the mineral nutrient boron. The cut stems and leaves exude a poisonous, white juice, which was a traditional remedy for warts. There are some 15 spurges in Britain and Ireland, several of them rare. This and Petty Spurge (*see* LOOKALIKES) are the only common spurges of cultivated land.

MILKWORT FAMILY, POLYGALACEAE

| J | F | M | A | M | J |
| J | A | S | O | N | D |

ID FACT FILE

Height:
10–35 cm

Flowers: 4–7 mm
long, with 5
petal-like sepals
and a tiny corol-
la, mostly blue
but also magen-
ta, lilac, pink,
white or white
flushed with blue;
10–40 in loose,
erect clusters

Leaves: Unpaired,
narrowly ellipti-
cal, the lower
ones smaller
than the upper

Fruits: Flattened,
egg-shaped cap-
sules c.5 mm
long, enclosed
within the green,
persistent calyx

Lookalikes:
Heath Milkwort
(*Polygala serpyl-
lifolia*), of acid
grassland, has at
least the lowest
leaves in oppo-
site pairs.

Common Milkwort
Polygala vulgaris

Slender, prostrate, sprawling or weakly ascending perennial; widespread in dry grassland,
especially on chalk or limestone soils, and on
sand-dunes and cliffs. The flowers, which are
intricate in structure, are variable in colour
even within a small area. This is the commonest of five native species of milkwort. Milk-
worts were prescribed by herbalists to nursing
mothers and were once a feature of garlands
for Rogation or 'beating the bounds' rituals.

MALLOW FAMILY, MALVACEAE

J F M A M J
J A S O N D

Common Mallow
Malva sylvestris

ID FACT FILE

HEIGHT:
20–100 cm,
sometimes up to
150 cm

FLOWERS: 2–4 cm
across, pinkish-
purple or lilac,
with darker veins,
with 5 deeply
notched petals

LEAVES: Long-
stalked, kidney- to
heart-shaped or
almost circular,
with 3–7 toothed
lobes; often with
a dark blotch

FRUITS: Disc-
shaped whorl of
1-seeded nutlets

LOOKALIKES: Musk
Mallow has
deeply cut leaves
and pale pink or
white flowers;
Dwarf Mallow
(*Malva neglecta*)
is smaller and
usually prostrate,
with pale lilac
flowers.

Prostrate, sprawling or erect, softly hairy biennial, or short-lived perennial, of waste places,
roadsides, dry and disturbed grassland and
hedge-banks. The plant adds colour to even
the scruffiest roadside. The leaves have soothing medicinal properties and have traditionally
been a source of winter greens in Europe and
elsewhere; they are still made into a soup in
parts of the Mediterranean region. The edible
fruits have long been known by children in
country districts as 'cheeses'.

Common, or Perforate, St John's Wort

Hypericum perforatum

ID FACT FILE

HEIGHT:
30–100 cm

FLOWERS:
10–15 mm across, yellow, in branched clusters; 5 petals, twisted and reddish in bud; 5 sepals, each with a few black dots; many yellow stamens

LEAVES: In opposite pairs, 1–2 cm long, oblong or oval, with many translucent dots

FRUITS: Pear-shaped capsules, 4–6 mm, splitting into 3 segments

LOOKALIKES: Slender St John's Wort (*Hypericum pulchrum*), a dainty plant up to 40 cm tall, with reddish petals, occurs on heaths.

Erect, hairless perennial of dry grasslands, scrub, woodland margins and waste ground; it is the commonest of the dozen or so native species of St John's wort, but rare in N Scotland. It was associated with Christian and pagan Summer Solstice festivals (St John's Day is 24 June) and was thought to ward off witchcraft and spells. It is poisonous to stock, making their skin sensitive to sunlight. It is a troublesome weed in N America, where it has been introduced.

VIOLET AND PANSY FAMILY, VIOLACEAE

J	F	M	A	M	J
J	A	S	O	N	D

Dog-violet
Viola riviniana

ID FACT FILE

HEIGHT: 5–20 cm

FLOWERS: 15–25 mm long, bluish-violet, with a pale spur, solitary on leafy stems

LEAVES: Long-stalked, heart-shaped, hairless

FRUITS: Pointed capsules, splitting into 3 segments

LOOKALIKES: Early Dog-violet (*Viola reichenbachiana*), mainly in S England, has smaller, all-violet flowers. Sweet Violet has white, purplish or violet, scented flowers.

A tufted perennial of hedge-banks, scrub, woodland, grassland, mountain ledges and coastal heaths. It can spread rapidly to become quite a persistent weed in gardens. Like all violets and pansies the flowers hang upside-down, as the flower-stalk bends sharply just below the flower. The largest petal is produced into a spur or sac. The seeds are dispersed by ants. There are several closely related dog-violets, which cross with one another to form a confusing range of hybrids.

VIOLET AND PANSY FAMILY, VIOLACEAE

Field Pansy
Viola arvensis

ID FACT FILE

HEIGHT: 5–40 cm

FLOWERS:
10–18 mm long, cream, variably marked with yellow and bluish-violet, solitary on leafy stems; petals shorter than the sepals

LEAVES: Oblong or spoon-shaped, lobed or toothed

FRUITS: Pointed capsules, splitting into 3 segments

LOOKALIKES: Wild Pansy or Heartsease (*Viola tricolor*) has larger, violet, purple or yellow flowers, the petals longer than the sepals.

An annual with weak, ascending stems; often abundant on cultivated land and one of the few weeds of arable land to have persisted in the face of modern chemical sprays. It is still widespread, except in N Scotland, and locally common in cereal crops. It is a very variable species that crosses with other wild pansies (see LOOKALIKES) to produce a complex array of forms; several of these variants have been described as species, subspecies and varieties.

CUCUMBER AND MARROW FAMILY, CUCURBITACEAE

J	F	M	A	M	J
J	A	S	O	N	D

White Bryony
Bryonia cretica

ID FACT FILE

Height: 1–4 m

Flowers: Greenish-white with darker veins, in stalked clusters, the male and female on different plants

Leaves: 5- or 7-lobed, like a large ivy leaf; long, spirally twisted tendrils opposite leaves

Fruits: Spherical red berries up to 1 cm across

Lookalikes: Black Bryony, another climbing plant of hedges, has hairless, heart-shaped leaves. Hop has cone- or tassel-like flowers.

Bristly, climbing or trailing perennial, arising from a stout, tuberous rhizome, growing in hedges and scrub and on woodland margins. The great root, thought fancifully to resemble a human body, has in the past been called Mandrake – and was once sold as such. (The true plant of that name grows in the Mediterranean region.) White Bryony is rarer in N and W Britain and absent from Ireland, except for one station near Dublin where it has been introduced. The plant is poisonous.

J	F	M	A	M	J
J	A	S	O	N	D

Purple Loosestrife
Lythrum salicaria

ID FACT FILE

Height: 50–180 cm

Flowers: 1 cm, reddish-purple, with 6 (sometimes 4) petals, in whorls, massed in long, dense spikes

Leaves: In opposite pairs or threes, spear-shaped or oval, clasping the stem slightly

Fruits: Egg-shaped capsules, 3–4 mm, with many tiny seeds

Lookalikes: Rosebay Willowherb has larger, 4-petalled flowers and grows in drier places.

Striking, erect, short-hairy perennial, with 4-angled stems, forming prominent clumps in marshes, beside rivers and streams, and on damp waste ground. In W Ireland especially it provides bold splashes of colour to the landscape during late summer, even away from water. Three types of flower are produced, on different plants, each with stigmas of different length; this modification enhances cross-pollination by insects. The sticky seeds are dispersed on the feet and feathers of waterfowl.

WILLOWHERB AND EVENING PRIMROSE FAMILY,
ONAGRACEAE

J	F	M	A	M	J
J	A	S	O	N	D

Evening Primrose

Oenothera biennis

ID FACT FILE

HEIGHT:
80–150 cm

FLOWERS: 4–6 cm
across, with 4
pale yellow
petals and green
sepals, in a long,
erect cluster

LEAVES: Short-
stalked, narrow,
broadly spear-
shaped,
shallowly toothed

FRUITS: Slender,
downy capsules
c.3 cm long

LOOKALIKES: Large-
flowered Evening
Primrose
(*Oenothera
erythrosepala*)
has hairs with
red, swollen
bases and
flowers 5–8 cm
across, with
reddish sepals.

Erect annual, or biennial, with downy stems;
locally common on roadsides and other open,
sandy ground, waste places and gardens. The
roots are edible and the plant is increasingly
being grown for its oil, which is used in
cosmetic and medical products. Although
introduced from N America, this plant has
become a familiar, established member of the
British flora; it is rarer in the north and in
Wales. The flowers open in the evening and
are pollinated by night-flying moths.

WILLOWHERB AND EVENING PRIMROSE FAMILY,
ONAGRACEAE

Rosebay Willowherb
or Fireweed
Chamerion angustifolium

J	F	M	A	M	J
J	A	S	O	N	D

ID FACT FILE

HEIGHT:
80–250 cm

FLOWERS:
c.25 mm across,
with 4 purplish-
pink petals; in
long, loose,
tapering spikes;
buds down-
turned

LEAVES: Alternate,
short-stalked,
narrow, spear-
shaped

FRUITS: Slender
capsules, up to
6 cm long,
splitting to
release the
numerous seeds,
each with a
plume of silky
hairs

LOOKALIKES:
Purple Loose-
strife has
smaller, usually
6-petalled flow-
ers and grows in
wet places.

Erect, almost hairless perennial, with creeping
rhizomes, forming extensive patches; a very
common and conspicuous plant of woodland
clearings, heathland, waste ground and derelict
industrial land, railway embankments and sand-
dunes. This species has expanded its range
greatly during the present century, especially
since World War II, when it spread rapidly in
the ruins of bombed buildings. However, it
remains uncommon over much of Ireland. The
hairy seeds are dispersed by the wind.

WILLOWHERB AND EVENING PRIMROSE FAMILY,
ONAGRACEAE

Great Hairy Willowherb or Codlins-and-Cream

Epilobium hirsutum

| J | F | M | A | M | J |
| J | A | S | O | N | D |

ID FACT FILE

Height: 1–2 m

Flowers: 20–25 mm across, in loose clusters, with 4 purplish-pink or occasionally white, shallowly notched petals, and a prominent, 4-lobed stigma; buds erect

Leaves: Opposite, half-clasping the stem, oblong or spear-shaped, shallowly toothed

Fruits: Slender, downy capsules, 4–8 cm long, splitting to release the numerous seeds, each with a plume of silky hairs

Lookalikes: Rosebay Willowherb has flowers in long spikes and grows in drier places.

Erect, little-branched, softly hairy perennial; a conspicuous and handsome plant of marshes, streamsides, riverbanks, waysides and waste places throughout Britain and Ireland, although absent from much of Scotland. It is the largest of the willowherbs and can form thickets that more or less block small streams. The commonly used local name of Codlins-and-Cream (codlin is an old word for cooking apple) may be a whimsical reference to the colour combination of rosy petals and cream stigma and stamens.

IVY FAMILY, ARALIACEAE

| J | F | M | A | M | J |
| J | A | S | O | N | D |

Ivy
Hedera helix

ID FACT FILE

HEIGHT: 1–5 m

FLOWERS:
Yellowish-green,
the parts in
fives, in erect,
stalked, domed
clusters

LEAVES: 5-lobed,
of characteristic
'ivy' shape;
those on the
flowering
branches
elliptical,
unlobed, pointed

FRUITS: Spherical,
flat-topped, black
berries, 6–8 mm
across

LOOKALIKES: An
unmistakable
plant, although
the leaves vary in
colour and
shape.

Familiar, dark green, woody climber, either trailing on the ground, climbing by means of dense, sucker-like roots or producing flowers on shrubby flowering stems. Common and widespread, Ivy is a prominent landscape feature, often carpeting the ground in woods or festooning trees, hedges and rocks, old buildings and walls. It is the last plant of the year to flower, attracting insects on sunny autumn days. Ivy is the main food plant of midsummer broods of caterpillars of the Holly Blue butterfly.

CARROT FAMILY, UMBELLIFERAE

J	F	M	A	M	J
J	A	S	O	N	D

Sea Holly
Eryngium maritimum

ID FACT FILE

HEIGHT:
20–60 cm

FLOWERS: Blue, many, in dense, spherical heads up to 3 cm across, arranged in a cluster amongst spiny bracts

LEAVES: Mostly 3-lobed, with wavy, spiny-toothed margins; margins and veins whitish; upper leaves clasping stem

FRUITS: Narrowly egg-shaped; spiny

LOOKALIKES: A number of similar spiny plants (eryngos) are grown in gardens.

A distinctive, stiff, spiny, bluish-green, hairless perennial, forming patches; locally common on sandy seashores, sometimes on shingle. It looks superficially like the thistles (daisy and dandelion family). The plant is well adapted to its dry, harsh habitat. Deep roots enable it to reach fresh water and the waxy leaf surfaces prevent excessive water loss in the dry conditions of the beach, as well as protecting against damage from salt spray. The roots were formerly candied and sold as a sweet.

CARROT FAMILY, UMBELLIFERAE

Cow Parsley or Queen Anne's Lace

Anthriscus sylvestris

J	F	M	A	M	J
J	A	S	O	N	D

ID FACT FILE

HEIGHT:
40–150 cm

FLOWERS: White,
3–4 mm across,
many, in dense,
flat-topped heads
or umbels
3–7 cm across;
a few oval bracts

LEAVES: Com-
pound, up to
30 cm long,
feathery, fern-like

FRUITS: Egg-
shaped, up to
1 cm long, flat-
tened, smooth,
dark brown or
black

LOOKALIKES:
Hemlock is taller
and less
branched, the
leaves are more
feathery and the
stems have
purple blotches.

An erect, rather robust, leafy, branched perennial; widespread and often growing in great crowds on roadsides, along hedgerows and in shadier places generally. This plant is the commonest and most familiar member of the carrot family, whitening the late spring landscape. The leaves begin to grow during winter and a few plants can be seen in flower as early as February. A variable number of plants have purplish stems and sometimes purplish leaves as well. The plant is poisonous.

CARROT FAMILY, UMBELLIFERAE

J	F	M	A	M	J
J	A	S	O	N	D

Wild Angelica
Angelica sylvestris

ID FACT FILE

HEIGHT:
60–180 cm

FLOWERS:
Petals to 15 mm long, white or pink, at first greenish. Flowering season July–August

LEAVES:
Large, twice-pinnate, with elliptic, toothed lobes. Bracts usually absent, Bracteoles numerous.

FRUITS:
Oval fruits with winged edges 4–6mm long.

LOOKALIKES:
Garden Angelica and Hemlock are similar.

Wild Angelica has a tall, robust, hollow-ridged stem, usually purplish tinged, and is a common plant found in damp places: wet meadows, marshes, fens, wet open woodlands, ditches and stream, river and lake margins. It prefers mildly acid to calcareous soils and is distributed throughout the whole of the British Isles. A perennial plant, Wild Angelica flowers from July to September.

CARROT FAMILY, UMBELLIFERAE

| J | F | M | A | M | J |
| J | A | S | O | N | D |

Ground Elder
Aegopodium podagraria

ID FACT FILE

HEIGHT:
30–100 cm

FLOWERS: White, many, in dense, flat-topped heads or umbels up to 6 cm across; no bracts

LEAVES: Stalked, divided 1–2 times into 3 oval or broadly spear-shaped, irregularly toothed leaflets

FRUITS: Narrowly egg-shaped, flattened, smooth, with 5 ridges

LOOKALIKES: The far-creeping habit distinguishes this familiar plant from other members of the family.

Hairless, far-creeping, rather aromatic perennial, forming large patches; common in damp, shady places, especially gardens, where it is a notorious weed. The plant spreads by means of shallowly rooted runners that are brittle and will regenerate readily to produce new plants. It is rarely found away from human habitation and is almost certainly introduced. The leaves were formerly cooked and eaten like spinach and the plant has a considerable reputation as a supposed remedy for gout.

CARROT FAMILY, UMBELLIFERAE

Wild Parsnip
Pastinaca sativa

ID FACT FILE

HEIGHT:
30–100 cm,
sometimes up to
150 cm

FLOWERS: Yellow,
many, in dense,
flat-topped heads
or umbels up to
10 cm across;
no bracts or
sometimes 1–2
that soon fall

LEAVES: Compound, with 5–11
broadly oblong,
toothed leaflets

FRUITS: Elliptical,
flat, narrowly
winged, 5–8 mm
long

LOOKALIKES: Fennel (*Foeniculum vulgare*), with
yellow flowers,
has feathery
leaves that smell
of aniseed; it
escapes from
gardens on to
waste ground
and dry banks.

A rough-hairy, often robust, erect, aromatic biennial of dry grassland, roadsides and waste ground, usually on lime-rich soils. It occurs northwards to N Yorkshire, but is widespread only south and east of a Severn–Humber line; it is local in Ireland, where it is not native. The plant is a subspecies of garden Parsnip, a taller and more robust plant with larger seeds that sometimes escapes on to road-verges. Parsnips were a major root vegetable before potatoes were introduced from North America.

CARROT FAMILY, UMBELLIFERAE

Hogweed
Heracleum sphondylium

J F M A M J
J A S O N D

ID FACT FILE

HEIGHT: 1–3 m

FLOWERS: White, or sometimes purplish-pink, many, in dense, flat-topped heads or umbels up to 25 cm across; a few, narrow bracts

LEAVES: Variably compound, rough-hairy

FRUITS: Oval, flat, broadly winged, 2–3 mm long

LOOKALIKES: Giant Hogweed (*Heracleum mantegazzianum*), introduced from the Caucasus, is a massive plant up to 5 m tall, with huge stems, leaves up to 3 m across and umbels 20–50 cm across.

A coarse, robust, hairy, erect perennial of roadsides, hedge-banks, lush grassland, streamsides, and woodland margins and clearings. Although now disregarded as just an unsightly weed, it was a traditional food for pigs, which were once kept by most rural households. Like several other members of the carrot family, notably Giant Hogweed (see LOOKALIKES), with which it sometimes crosses, the sap causes the skin to become sensitive to sunlight, resulting in unpleasant blisters and soreness.

CARROT FAMILY, UMBELLIFERAE

J	F	M	A	M	J
J	A	S	O	N	D

Wild Carrot
Daucus carota

ID FACT FILE

HEIGHT:
20–100 cm

FLOWERS: White or lilac, numerous, in dense, flat-topped, concave or domed heads or umbels 2–6 cm across; ruff of many deeply divided bracts

LEAVES: Compound, deeply divided, fern-like

FRUITS: Egg-shaped, 2–4mm long, flattened, densely spiny

LOOKALIKES: The ruff of deeply divided bracts gives this plant a distinctive appearance compared with other members of the family.

Annual, biennial or short-lived perennial, with a long, narrow root and erect stems; a widespread, if local, plant of dry grasslands, especially on chalk or limestone and by the sea. The concave heads of fruits are a distinctive feature of roadsides and grassland in summer. A very variable species; many plants on the coast are not more than 40 cm tall and have fleshy, shiny leaves and flat heads of fruits. Wild Carrot is a subspecies of the garden carrot, but lacks its swollen, orange root.

HEATH FAMILY, ERICACEAE

J	F	M	A	M	J
J	A	S	O	N	D

Heather or Ling

Calluna vulgaris

ID FACT FILE

HEIGHT:
20–60 cm,
sometimes up to
100 cm

FLOWERS: Tiny,
bell-shaped, pale
purple, rarely
lilac or white, in
leafy spikes;
corolla and calyx
4-lobed

LEAVES: Small,
overlapping in
opposite rows,
narrowly oblong

FRUITS: Small,
spherical cap-
sules, each
enclosed by dry,
persistent corolla

LOOKALIKES: Other
heathers have
larger flowers
and non-overlap-
ping leaves.

Evergreen, branched shrublet, often rather woody below; widespread and often dominant over huge areas of moors, heaths, bogs, open woods, grasslands and sand-dunes. Plants are sometimes grey-downy, especially on coastal heaths. In Scotland and elsewhere heather formerly provided thatch and bedding and was used as a source of a dye and to make a legendary beer. It is still an important source of nectar for bees. Regular burning and grazing prevents heather from becoming too overgrown.

HEATH FAMILY, ERICACEAE

Bilberry, Whortleberry or Blaeberry
Vaccinium myrtillus

J	F	M	A	M	J
J	A	S	O	N	D

ID FACT FILE

HEIGHT:
20–40 cm

FLOWERS:
Solitary or
paired, 4–6 mm
long, sac-like,
almost spherical,
pale green tinged
pink

LEAVES: Oval,
pointed, minutely
toothed, bright
green, falling in
autumn

FRUITS: Spherical,
bluish-black
berries 5–8 mm
across

LOOKALIKES:
Related species
of more local
distribution, with
black or red
berries, are
found on bogs,
moors and
mountains.

Little-branched, hairless shrublet of moors, heaths and open woods of birch, oak and pine on well-drained acid soils; often very abundant, but scarce in C England and E Anglia and much of the midlands of Ireland. The edible, sweet berries were formerly gathered commercially by country people, but are now encountered only rarely in shops and restaurants in Britain and Ireland. They are still extensively collected in C and E Europe, and are excellent eaten raw or cooked in tarts, jelly or jam. They also make a purplish dye.

PRIMROSE FAMILY, PRIMULACEAE

J	F	M	A	M	J
J	A	S	O	N	D

Oxlip
Primula elatior

ID FACT FILE

HEIGHT
15–30 cm

FLOWERS
Pale yellow,
5–25 mm in
diameter, not
fragrant, 1–20 in
a nodding one-
sided cluster

LEAVES
Abruptly
narrowed at
base. All in a
basal rosette,
more or less
oval, abruptly
contracted into
stalk, indistinctly
toothed, wrinkled

FRUITS
Capsules
enclosed within
persistent calyx-
tube

LOOKALIKES
Leaves of False
Oxlip are more
gradually tapered
to the base than
Oxlip's, the flow-
ers are a deeper
yellow with folds
in the throat

This low, hairy perennial is found in woods, scrub, grassland, and usually in a considerable quantity in Eastern England. It prefers the poorly-drained soils of the chalky boulder clay found in East Anglia, and is most present in areas where the primrose is scarce. The Oxlip is a perennial and flowers during April and May. It is often confused with False Oxlip, a hybrid of Cowslip and Primrose (see LOOKALIKES).

PRIMROSE FAMILY, PRIMULACEAE

J	F	M	A	M	J
J	A	S	O	N	D

Yellow Loosestrife
Lysimachia vulgaris

ID FACT FILE

HEIGHT:
40–160 cm

FLOWERS:
Cup-shaped,
15–18 mm
across, with
5 petals and
5 sepals, clus-
tered in loose,
branched heads

LEAVES: In oppo-
site pairs or in
threes or fours,
short-stalked,
oval to spear-
shaped, with
black dots

FRUITS: Spherical
capsules

LOOKALIKES: Dot-
ted Loosestrife
(*Lysimachia
punctata*) is a
garden escape
with flowers
2–3 cm across
in long, dense
clusters.

An erect perennial, spreading by rhizomes and forming clumps in marshes, wet woods, shores of lakes and gravel-pits, riverbanks and near houses as an escape from gardens. It is wide-spread, if rather local, throughout Britain and Ireland, except for much of N and E Scotland. This plant is not related to Purple Loosestrife, although they occur in similar marshland and wet habitats. 'Loosestrife' may refer to a calm-ing effect that these plants supposedly have on livestock.

PRIMROSE FAMILY, PRIMULACEAE

Scarlet Pimpernel
Anagallis arvensis

ID FACT FILE

Height: 5–50 cm

Flowers: In pairs, arising from leaf-axils, dish-shaped, 4–8 mm across, scarlet, with a purple centre, sometimes flesh-coloured (rarely blue or lilac)

Leaves: In opposite pairs, short-stalked, oval to spear-shaped, sometimes slightly fleshy

Fruits: Spherical capsules c.5 mm across, splitting round the middle

Lookalikes: Yellow Pimpernel is perennial, with slightly larger, yellow flowers, and grows in damp, shady places.

A prostrate, hairless, ascending or weakly erect annual or biennial, with square stems. It is a common plant of cultivated land, waysides, sand-dunes and open, damp or sandy places near the sea, and the prettiest of our common weeds. The flowers open only in bright sunshine – hence its other name, 'Poor Man's Weatherglass' – although they close in late afternoon as well. Plants with flesh-coloured flowers are locally common on the coasts of Ireland, SW England and elsewhere.

SEA-LAVENDER FAMILY, PLUMBAGINACEAE

| J | F | M | A | M | J |
| J | A | S | O | N | D |

Common Sea-lavender
Limonium vulgare

ID FACT FILE

HEIGHT:
20–40 cm, sometimes up to 70 cm

FLOWERS: 5–6 mm across, funnel-shaped, bluish-lilac, in a dense, 1-sided, flat-topped, branched cluster

LEAVES: All basal, stalked, spear-shaped or elliptical, more or less pointed, fleshy

FRUITS: Small, 1-seeded, dry, papery

LOOKALIKES:
Lax-flowered Sea-lavender (*Limonium humile*), with stems up to 40 cm tall, branched in upper half, and flowers in longer, looser spikes, has a similar distribution but also occurs in Ireland.

Erect, hairless perennial, with rather woody, branched rhizomes and leafless stems, a conspicuous feature of muddy saltmarshes and sometimes coastal cliffs, rocks and shingle beaches. It is widespread and often common on coasts north to the Firth of Forth, although absent from Ireland (but see Lax-flowered Sea-lavender, LOOKALIKES). In July and August, it tints saltmarshes with its distinctive, lavender-like colour. The dried flowers are sometimes used in flower-arranging.

GENTIAN FAMILY, GENTIANACEAE

J	F	M	A	M	J
J	A	S	O	N	D

Centaury
Centaurium erythraea

ID FACT FILE

HEIGHT:
10–30 cm, sometimes up to 50 cm

FLOWERS: Pink, funnel-shaped, 5–8 mm across in a dense, branched, more or less flat-topped cluster

LEAVES: Basal leaves oval, stem leaves much smaller and narrower

FRUITS: Small, cylindrical capsules

LOOKALIKES: There are a number of other, rarer, pink-flowered species of centaury, all of them strictly coastal.

Erect, hairless biennial or short-lived perennial, usually with a single stem, of dry grassland, sea-cliffs and sand-dunes. This attractive wild flower is our commonest native gentian relative; it is generally widespread, but it is local and mostly coastal in Scotland. It is a variable species, especially where it grows on the coast – here, dwarf and narrow-leaved variants occur. The plant has healing properties and an infusion made from it has long been used as a tonic to aid digestion.

BEDSTRAW AND MADDER FAMILY, RUBIACEAE

| J | F | M | A | M | J |
| J | A | S | O | N | D |

Lady's Bedstraw
Galium verum

ID FACT FILE

HEIGHT:
10–120 cm

FLOWERS: Yellow, scented, 2–3.5 mm across, with 4 spreading petal-lobes, in long, dense, often branched clusters

LEAVES: Very narrow, 1-veined, sharp-pointed, in whorls of 8–12

FRUITS: 2-lobed, smooth, 1–1.5 mm long

LOOKALIKES: Squinancywort (*Asperula cynanchica*) is a larger perennial plant, with 2–4 leaves in a whorl, and pink or white flowers. Hedge Bedstraw has white flowers.

Attractive, finely hairy, erect or sprawling perennial, with creeping, underground stems, a widespread plant of dry grasslands, hedge-banks, sand-dunes, shingle beaches, sometimes in churchyards and on lawns, especially on lime-rich soils. Hay made from this useful plant was popular for stuffing mattresses, not only for its sweet scent but also because it deters fleas and other vermin. Lady's Bedstraw was formerly used as a substitute for rennet to curdle cheese. The underground stems yield a red dye.

BEDSTRAW AND MADDER FAMILY, RUBIACEAE

J	F	M	A	M	J
J	A	S	O	N	D

Hedge Bedstraw
Galium mollugo

ID FACT FILE

HEIGHT:
30–160 cm

FLOWERS: White,
2–3 mm across,
with 4 spreading
petal-lobes, in a
lax, much-
branched cluster

LEAVES: Narrow,
oblong, pointed,
with rough-prickly
margins, in
whorls of 6–8

FRUITS: 2-lobed,
rough, purplish
or greyish,
1–2 mm long

LOOKALIKES:
Heath Bedstraw
(*Galium saxatile*),
a plant of heaths
and pastures on
poor soils, is not
more than 35 cm
tall, with mostly
prostrate stems
and leaves in
whorls of 4–6.

Erect or spreading perennial, with long, under-ground runners and often hairy, but not prick-ly, stems; a widespread plant of dry grassland, woodland clearings and hedges. The perennial bedstraws, plants of grasslands and marshes, and all except Lady's Bedstraw with white flowers, are a difficult group of plants to distin-guish from one another. Occasional hybrids with Lady's Bedstraw, which occur where the two grow together and cross, have pale yellow flowers.

BEDSTRAW AND MADDER FAMILY, RUBIACEAE

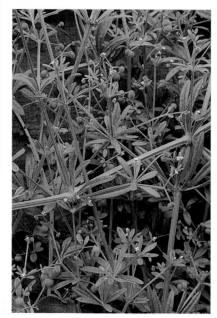

| J | F | M | A | M | J |
| J | A | S | O | N | D |

Goosegrass or Cleavers
Galium aparine

ID FACT FILE

Height:
50–180 cm

Flowers: Greenish-white or white, 1.5–2 mm across, with 4 spreading petal-lobes, in loose, fewflowered clusters

Leaves: Narrow, oblong, pointed, prickly with hooked bristles, in whorls of 6–9

Fruits: 2-lobed, densely covered with hooked bristles, 3–6 mm long

Lookalikes: Hedge Bedstraw is perennial with smooth rather than bristly stems and dense clusters of white flowers.

A scrambling or climbing, bristly annual, with 4-angled stems, of scrub, woodland margins, hedge-banks, cultivated land and shingle beaches. The bristly fruits cling readily to clothes and animal fur – one rarely returns from a country walk without a few in one's socks or the dog's coat – effectively dispersing the seeds. Indeed, the whole plant will stick to clothes like 'velcro', demonstrating how the bristles are its means of support. A variable species: plants from cultivated ground often have smaller fruits.

BINDWEED FAMILY, CONVOLVULACEAE

Field Bindweed
Convolvulus arvensis

ID FACT FILE

HEIGHT:
50–200 cm

FLOWERS: White, or pink variously striped with white, funnel-shaped, 15–35 mm across, slightly scented, 1–3 on stalks about as long as the leaves

LEAVES: Oblong, triangular, spear- or arrow-shaped

FRUITS: Almost spherical, 2-celled capsules

LOOKALIKES: Hedge Bindweed is a larger plant with usually white flowers 3–6 cm across.

Slender, elegant, often hairy perennial, arising from an extensive, branched root system, with numerous twining (anti-clockwise) stems, which exude a white milky juice when cut. This is a familiar and abundant plant of cultivated and waste ground, also amongst grass, either prostrate or climbing other plants and wire-netting fences. Even quite small fragments of the deep roots can grow into new plants. The flowers close during dull or wet weather and in late afternoon. A very variable species.

BINDWEED FAMILY, CONVOLVULACEAE

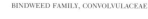

Hedge Bindweed
Calystegia sepium

| J | F | M | A | M | J |
| J | A | S | O | N | D |

ID FACT FILE

HEIGHT: 1–3 m

FLOWERS: White, sometimes pink or white, funnel-shaped, 3–6 cm across, unscented; 2 bracts, sometimes inflated, at base

LEAVES: Heart- or arrow-shaped

FRUITS: Spherical, 1-celled capsules, enclosed by persistent sepals

LOOKALIKES: Field Bindweed is a smaller plant with white or pink flowers 15–35 mm across; Sea Bindweed (*Calystegia soldanella*), with short stems, round leaves and pink flowers, occurs on sand-dunes.

A perennial climber, with tough, twining stems, that exude a white, milky juice when cut; a frequent plant of wood margins, scrub, marshes and waste ground, too often a serious garden weed. Sometimes hedges, banks, wire-netting fences and the wires supporting telegraph poles are festooned with a dense growth of this plant. A variable species; an introduced subspecies has more vigorous leafy growth and larger flowers 6–8 cm across, with strongly inflated bracts at the base.

COMFREY AND FORGET-ME-NOT FAMILY,
BORAGINACEAE

J	F	M	A	M	J
J	A	S	O	N	D

Viper's Bugloss
Echium vulgare

ID FACT FILE

HEIGHT:
30–100 cm

FLOWERS: Blue,
with purplish-red
buds, tubular,
1–2 cm long, in
curved, 1-sided
clusters forming
a tall spike;
4–5 stamens,
the longer ones
protruding

LEAVES: Elliptical
or strap-shaped,
the upper ones
narrower

FRUITS: 4 nutlets

LOOKALIKES:
Borage (*Borago
officinalis*) has
oval or spear-
shaped leaves
and flowers with
5 widely spread-
ing segments, in
loose clusters; it
is a garden
escape.

Bristly, erect annual or biennial of dry grassland, waysides, roadside rubble, bare chalk or
gravel, and shingle beaches. Widespread but
local in England, it is rarer in Scotland and
occurs in Ireland mainly near the E coast. The
flowers are very striking, with their purplish-
red buds; in quantity they can be a wonderful
show of blue, not a common colour amongst
our native wild flowers. This plant causes no
trouble in Britain, but is a major weed in parts
of the USA where it has been introduced.

COMFREY AND FORGET-ME-NOT FAMILY,
BORAGINACEAE

J	F	M	A	M	J
J	A	S	O	N	D

Comfrey
Symphytum officinale

ID FACT FILE

HEIGHT:
50–120 cm

FLOWERS: White or violet-purple, pinkish or blue, tubular, 12–18 mm long, in 1-sided clusters

LEAVES: Broadly spear-shaped, pointed, the upper ones stalkless

FRUITS: 4 smooth, black, shiny nutlets

LOOKALIKES: The most widespread of several comfreys, mostly garden escapes. Tuberous Comfrey (*Symphytum tuberosum*), up to 40 cm tall with pale yellow flowers, is common in woods in N Britain.

Robust, erect, branched, rough-hairy perennial, with winged stems, forming large, untidy patches in hedgerows, waysides, ditches, damp grassland and the banks of streams and rivers. This is a most vigorous and valuable plant; in recent years it and related species have assumed almost a cult status amongst herbal healers and gardeners. Comfrey reduces inflammation and aids healing; the leaves can be eaten like spinach or fried in batter; and the whole plant makes a mineral-rich garden compost.

COMFREY AND FORGET-ME-NOT FAMILY,
BORAGINACEAE

Field Forget-me-not
Myosotis arvensis

ID FACT FILE

HEIGHT: 5–25 cm

FLOWERS: Pale blue, with a yellow eye, saucer-shaped, 3–5 mm across, in leafless, 1-sided, rather flat-topped clusters

LEAVES: Elliptical, the upper ones smaller, spear-shaped

FRUITS: 4 nutlets, enclosed within calyx

LOOKALIKES: Two other forget-me-nots of dry, open ground have flowers 2–3 mm across: Early Forget-me-not (*Myosotis ramosissima*) has blue flowers; Changing Forget-me-not (*Myosotis discolor*) has flowers that open yellow and change to blue.

Erect annual, of open and disturbed ground, especially arable fields. The flowers of the forget-me-nots demonstrate well the character-istic structure found in the comfrey and forget-me-not family. The clusters are short and condensed in bud, but the main stalk lengthens and curves as the flowers open to produce a shape reminiscent of the curled tail of the scorpion. From this derives an older name for Forget-me-not, Scorpion-grass, which comes to us from both French and German.

COMFREY AND FORGET-ME-NOT FAMILY,
BORAGINACEAE

Water Forget-me-not
Myosotis scorpioides

ID FACT FILE

HEIGHT:
10–40 cm

FLOWERS: Saucer-shaped, pale blue, with a yellow eye, rarely pink or white, 8–10 mm across, in leafless, 1-sided, flat-topped clusters; 5 triangular calyx-teeth

LEAVES: In opposite pairs, spear- or spoon-shaped, blunt, the upper ones smaller

FRUITS: 4 black, shiny nutlets, enclosed in the calyx

LOOKALIKES: The short, triangular calyx-teeth distinguish this plant from other species of forget-me-not of wet places.

Erect, hairless to slightly hairy, pale green perennial with creeping stolons or runners; widespread and often abundant beside and in streams, in marshes and wet woods, especially rides and glades, and damp meadows. This plant can be a fine sight when in flower. Other similar forget-me-nots occur in wet places: for example, Creeping Forget-me-not (*Myosotis secunda*), of acid marshes and bogs, especially in N and W Britain and Ireland, which is hairier and has flowers 6–8 mm across.

MINT FAMILY, LABIATAE

J	F	M	A	M	J
J	A	S	O	N	D

Bugle
Ajuga reptans

ID FACT FILE

HEIGHT:
10–40 cm

FLOWERS: Blue, rarely pink or white, 14–18 mm, in clusters amongst leafy bracts, tubular, 1–2 cm long, in a tall spike

LEAVES: In opposite pairs, oval, stalked, obscurely toothed, the upper ones smaller

FRUITS: 4 nutlets

LOOKALIKES: The blue flowers in tall spikes distinguish Bugle from other members of the mint family; Skullcap has flowers in pairs in a loose, 1-sided spike.

Erect perennial with long runners, the stems hairy on opposite sides at each leaf pair; widespread and often common in woods, especially rides and glades, hedges and damp meadows. The upper stems and leaves have something of the blue tint of the flowers. The plant has had a wide range of herbal uses, notably to staunch bleeding. Long established as a cottage garden plant, it remains popular as a cover plant for rockeries and borders, especially variants with bronze or multicoloured leaves.

MINT FAMILY, LABIATAE

J	F	M	A	M	J
J	A	S	O	N	D

Wood Sage
Teucrium scorodonia

ID FACT FILE

HEIGHT:
20–50 cm

FLOWERS: Greenish-yellow, sometimes white or marked with red, 8–10 mm long, in pairs grouped in loose, spikelike clusters; hairy calyx

LEAVES: In opposite pairs, triangular to oval, with heartshaped base

FRUITS: 4 nutlets

LOOKALIKES: Yellow Archangel has a creeping rather than shrubby habit and has larger flowers in late spring.

Erect, branched, rather dowdy shrublet of heaths, scrub, dry, sandy or limestone banks, and rocky ground. It grows on both lime-rich and acid soils; research has shown that the adaptation of plants to grow on either soil type has a genetically controlled basis. The flowers appear dull, but they are neatly structured and sometimes attractively and richly marked with red. The plant was formerly used to flavour and preserve beer and has been used as a medicinal herb with healing properties.

MINT FAMILY, LABIATAE

Skullcap
Scutellaria galericulata

ID FACT FILE

HEIGHT:
10–50 cm

FLOWERS: Blue, with whitish-spotted lip, 1–2 cm long, in pairs in a loose, 1-sided spike

LEAVES: In opposite pairs, ovate or broadly spear-shaped, with a few round teeth

FRUITS: 4 nutlets

LOOKALIKES: Bugle has more numerous blue flowers clustered in a denser spike.

Erect, downy perennial, with creeping, rooting runners, of marshes, banks and margins of streams and rivers, damp woods and meadows. It is a widespread and sometimes common plant, but is local in E Scotland and in Ireland. Lesser Skullcap (*Scutellaria minor*), a smaller plant up to 15 cm tall, with more or less tooth-less leaves and purple-spotted, lilac flowers, is found on damp heaths and moors, mostly in the west. It sometimes crosses with Skullcap to form hybrids.

MINT FAMILY, LABIATAE

| J | F | M | A | M | J |
| J | A | S | O | N | D |

Hemp-nettle
Galeopsis tetrahit

ID FACT FILE

HEIGHT:
10–50 cm

FLOWERS: Pink or whitish, sometimes pale yellow, with darker markings, 15–20 mm long, in dense whorls; bristly calyx

LEAVES: In opposite pairs, oval or broadly spear-shaped, pointed, coarsely toothed

FRUITS: 4 nutlets

LOOKALIKES: Red Dead-nettle is a smaller plant with pinkish-purple flowers and pointed but not bristly calyx-teeth.

Coarsely hairy annual, the square stems with hairs on opposite sides; a widespread but rather local plant of open woodland, heaths, marshes and cultivated land. It is much less common as an arable weed than formerly. This plant is famous amongst plant evolutionists as the first species to be recreated in the botanic garden by crossing, followed by doubling up of the genetic material of the hybrid, of the two related species from which it was thought to have been derived in the wild. There are five similar species of hemp-nettle, mostly uncommon.

MINT FAMILY, LABIATAE

J	F	M	A	M	J
J	A	S	O	N	D

White Dead-nettle
Lamium album

ID FACT FILE

HEIGHT:
20–80 cm

FLOWERS:
Creamy white,
18–25 mm long,
in conspicuous,
compact whorls;
upper lip of corol-
la domed; bristle-
like calyx-teeth

LEAVES: In oppo-
site pairs, trian-
gular to oval,
pointed, coarsely
toothed

FRUITS: 4 nutlets

LOOKALIKES:
Yellow Archangel
has yellow
flowers.

Hairy, erect perennial of roadsides, hedges, shady waste ground and gardens. It occurs throughout Britain but is rare or local in N and W Scotland and in W and SW Ireland, where it is probably introduced. It rarely occurs very far from buildings, roads or paths. One of the plant's local names is Adam-and-Eve-in-the-Bower, alluding to the pair of black and yellow stamens that lie side by side in the domed upper lip of the flower. The flowers are attractive to bumblebees. Dead-nettles lack stinging hairs and are not related to Stinging Nettle.

MINT FAMILY, LABIATAE

ID FACT FILE

HEIGHT:
10–40 cm

FLOWERS: Pinkish-purple, rarely pale pink or white, 10–18 mm long, in conspicuous whorls; calyx-teeth pointed, not bristly

LEAVES: In opposite pairs, oval, heart-shaped at the base, pointed, toothed

FRUITS: 4 nutlets

LOOKALIKES: Hemp-nettle is a larger, coarser plant with pink flowers and bristle-like calyx-teeth.

Red Dead-nettle
Lamium purpureum

Downy, ascending or spreading annual, aromatic when crushed, the whole plant often purple-tinged; ubiquitous on cultivated and waste ground and a familiar garden weed. The flowers are very attractive to bumblebees. Red Dead-nettle is one of the first flowers to appear in late winter and early spring, along with Chickweed, Shepherd's Purse and Groundsel. Dead-nettles do not have stinging hairs and are not at all related to Stinging Nettle.

MINT FAMILY, LABIATAE

J	F	M	A	M	J
J	A	S	O	N	D

Marsh Woundwort
Stachys palustris

ID FACT FILE

HEIGHT:
20–120 cm

FLOWERS: Pinkish-purple,
12–15 mm long,
in whorls, forming a dense,
pyramidal spike

LEAVES: In opposite pairs,
spear-shaped,
short-stalked or
stalkless,
coarsely toothed

FRUITS: 4 nutlets

LOOKALIKES:
Hedge Woundwort has a
stronger smell
when bruised,
solid stems,
stalked leaves
and darker red
flowers.

An erect, creeping perennial, with tuberous rhizomes and hollow stems; widespread in marshes, the margins of lakes, ponds and rivers, ditches and damp fields, and as a weed; in Ireland it is less restricted to damp ground and is frequently a weed of disturbed and cultivated ground and pastures. It has been used, like several other plants in this family, to staunch bleeding. It has a smell when bruised, but not nearly as strong and unpleasant as that of Hedge Woundwort.

MINT FAMILY, LABIATAE

| J | F | M | A | M | J |
| J | A | S | O | N | D |

Hedge Woundwort
Stachys sylvatica

ID FACT FILE

HEIGHT:
50–120 cm

FLOWERS: Reddish-purple, with whitish blotches, rarely pink or white. 13–18 mm long, in whorls, forming a dense, pyramidal spike

LEAVES: In opposite pairs, oval, stalked, pointed, coarsely toothed

FRUITS: 4 nutlets

LOOKALIKES:
Marsh Woundwort has solid stems, narrower, unstalked leaves and more pinkish flowers; it is a plant of less shady places.

Erect, roughly hairy, unpleasant-smelling perennial, with a creeping rhizome and solid stems; a widespread and common plant of woods, hedges, shady places, abandoned cultivated land and overgrown gardens. The whole plant has a strong smell when bruised or damaged. Nevertheless, it has been used, like several other plants in this family, to staunch bleeding and to heal wounds. This is a characteristic flower of shady places during summer's heat.

MINT FAMILY, LABIATAE

J F M A M J J A S O N D

Ground-ivy
Glechoma hederacea

ID FACT FILE

HEIGHT:
10–50 cm

FLOWERS: Deep
violet, dark-spot-
ted, rarely pink
or white,
15–25 mm long,
3–6 in whorls

LEAVES: In oppo-
site pairs, heart-
or kidney-
shaped, stalked,
with scalloped
margins

FRUITS: 4 nutlets

LOOKALIKES:
Self-heal has
bluish-purple
flowers, and
blooms later in
the year.

Rather slender, softly hairy perennial, sometimes the whole plant tinged purple, with
far-creeping, rooting stems; often abundant in
woods, hedges and shady banks, also in
churchyards, untended gardens and grassy
places. Some plants have female as well as hermaphrodite flowers, to encourage cross-pollination. A bitter-tasting plant that was formerly
used medicinally and to flavour and preserve
beer. This is one of the first flowers of spring,
but one which is at its best well into May.

MINT FAMILY, LABIATAE

J	F	M	A	M	J
J	A	S	O	N	D

Self-heal
Prunella vulgaris

ID FACT FILE

HEIGHT: 5–50 cm

FLOWERS: Rich bluish-purple, sometimes violet, pink or white, 12–15 mm long, in a dense, cylindrical spike; purplish bracts

LEAVES: In opposite pairs, oval, stalked, untoothed or slightly toothed

FRUITS: 4 nutlets

LOOKALIKES: Ground-ivy has violet flowers, and blooms earlier in the year; Bugle and Skullcap have blue flowers.

Hairy, spreading or ascending annual, biennial or short-lived perennial; common in woods, waste places and waysides, damp or dry grassland, on sand-dunes and often in lawns. A very variable species; dwarfed, prostrate plants from lawns keep their features even when grown in good garden soil. The spikes of persistent purplish bracts are distinctive even when the flowers have fallen. The common name reflects a long and useful history of herbal use for staunching and healing wounds.

MINT FAMILY, LABIATAE

J	F	M	A	M	J
J	A	S	O	N	D

Large Thyme
Thymus pulegioides

ID FACT FILE

HEIGHT
2–30 cm

FLOWERS
Pale purple
colour, elongat-
ed, in whorled
spikes, 6mm in
length. Flowers
from June to
October

LEAVES
The leaves are
oval to spoon-
shaped and nar-
row towards the
stalk, with long
bristles at the
base

FRUIT
4 nutlets

LOOKALIKES
Similar to Breck-
land Thyme,
which has long,
creeping rounded
stems that are
hairy on all sides
and narrow
lanceolate
leaves.

An aromatic undershrub, which is low or short,
Large Thyme has no runners. The stems are
always hairy all round. Large Thyme is found
throughout southern England and as far north as
Yorkshire. It is rare in Wales, Ireland and the
south west and is present on dry grassland, well-
drained south-facing slopes and usually on chalk.
Like other species of Thyme, it has separate
female and hermaphrodite flowers on different
plants and is mainly pollinated by bees. The
female flowers are smaller.

MINT FAMILY, LABIATAE

Gipsywort
Lycopus europaeus

ID FACT FILE

HEIGHT:
30–100 cm

FLOWERS: Bell-shaped, 3–4 mm long, whitish with purple dots, in wide-spaced whorls; long, hairy calyx; stamens protruding

LEAVES: In opposite pairs, spear-shaped, short-stalked, jaggedly toothed, without a smell when bruised

FRUITS: 4 nutlets

LOOKALIKES: Water Mint and Corn Mint (*Mentha arvensis*) have larger, lilac flowers; those of Water Mint are mostly in a terminal head, and the leaves are strongly aromatic.

Erect, hairy, mint-like perennial, forming patches by means of a creeping rhizome and runners; a widespread, common and rather conspicuous plant of ditches, margins of ponds, streams, rivers and canals. It is more local in much of N Britain and Ireland. The neatly tiered pairs of opposite leaves give this plant a characteristic appearance. It yields a black dye, the basis of an old and scurrilous rumour of its use by gipsies to dye their hair, to enhance their exotic appearance!

MINT FAMILY, LABIATAE

J	F	M	A	M	J
J	A	S	O	N	D

Water Mint
Mentha aquatica

ID FACT FILE

HEIGHT:
20–80 cm

FLOWERS: Lilac,
3–4 mm long,
most of them in
a terminal head,
others in lower,
spaced whorls;
long, hairy calyx;
stamens
protruding

LEAVES: In oppo-
site pairs, oval,
stalked, toothed

FRUITS: 4 nutlets

LOOKALIKES: Corn
Mint (*Mentha
arvensis*) has
paler lilac
flowers, all in
separate whorls
up the stem; it
grows in drier
places.

Richly aromatic, hairy, erect perennial of
marshes, ditches and the sides of rivers, ponds
and lakes. This is the commonest mint of wet
places, giving marsh vegetation its characteristic odour in summer and autumn. Garden
Peppermint, familiar as the principal ingredient of mint sauce, is a hybrid between Water
Mint and garden Spearmint. These and other
mints are vigorous plants that frequently
escape from gardens to become naturalised in
damp or waste places.

J	F	M	A	M	J
J	A	S	O	N	D

Black Nightshade
Solanum nigrum

Erect or ascending, branched and often bushy, hairless or downy annual, of arable fields and gardens; widespread in England and Wales but rare in N Britain and Ireland. The lustrous-looking berries, which ripen from green to black during August to October, are poisonous – like those of its cultivated relative, the potato. Although the leaves, too, contain variable amounts of poisonous alkaloids, in S Europe they are sold, cooked and eaten as a green vegetable similar to spinach.

ID FACT FILE

HEIGHT:
10–60 cm

FLOWERS: White, 5–8 mm across, the petals becoming down-curved, in small, loose clusters; stamens in yellow cone

LEAVES: 3–6 cm, stalked, more or less oval or diamond-shaped, usually irregularly toothed, pointed

FRUITS: Loose cluster of spherical, shiny black berries

LOOKALIKES: Deadly Nightshade (*Atropa belladonna*) is a robust perennial, with bell-shaped flowers 2–3 cm long and berries up to 2 cm across, local in woods and scrub on chalk and limestone.

Woody Nightshade
Solanum dulcamara

ID FACT FILE

HEIGHT:
50–200 cm,
sometimes up to
400 cm

FLOWERS: Violet,
10–15 mm
across, in loose
clusters; each
petal with 2 green
nectary patches
at base; stamens
in a yellow cone

LEAVES: 5–8 cm,
the lower 3-lobed,
the upper spear-
shaped

FRUITS: Broadly
egg-shaped, scar-
let, shiny, translu-
cent berries,
c.1 cm long

LOOKALIKES: Dead-
ly Nightshade
(*Atropa bella-
donna*) is more
robust, with bell-
shaped flowers
and larger, black
berries; local in
woods and scrub
on chalk and
limestone.

Rather woody, climbing or straggling, downy
perennial, of shady places by streams and
rivers, hedgerow ditches, overgrown gardens
and damp woods. It is a widespread plant, but
in Scotland it occurs mainly on the coast and
along rivers. It is a characteristic species of
swamp woodland, a rare habitat that survives in
parts of East Anglia and Ireland; a prostrate,
fleshy variant grows on coastal shingle beaches.
The plant spreads both by seed and by the
production of new shoots from the roots.

FIGWORT AND FOXGLOVE FAMILY, SCROPHULARIACEAE

Great Mullein or Aaron's Rod

Verbascum thapsus

J	F	M	A	M	J
J	A	S	O	N	D

ID FACT FILE

HEIGHT:
80–200 cm

FLOWERS: Pale yellow, 20–50 mm across, in small clusters in the angle of a bract; massed in huge, unbranched spikes

LEAVES: Oval or broadly spear-shaped, pointed

FRUITS: Egg-shaped capsules containing numerous tiny seeds

LOOKALIKES: Several other species of mullein occur in similar habitats in Britain, mainly in the south, but not in Ireland.

Conspicuous, robust, erect, white-felted biennial, of sunny banks, waste ground, dry roadsides, scrub and wood margins; in Ireland it is often associated with houses or ruins. It is generally a widespread plant, sometimes occurring in crowds, but is rare over much of Ireland and in W and N Scotland. Like other biennials, the young plant develops a rosette of leaves at the end of the first summer, which looks rather like a hairy cabbage. The adult plant can produce many thousands of seeds.

FIGWORT AND FOXGLOVE FAMILY, SCROPHULARIACEAE

J	F	M	A	M	J
J	A	S	O	N	D

Common Figwort
Scrophularia nodosa

ID FACT FILE

HEIGHT:
40–100 cm

FLOWERS: Helmet-like, 10 mm long, green with a purplish-brown upper lip, in loose, leafy clusters

LEAVES: In opposite pairs, oval, short-stalked with cut-off base, double-toothed, pointed

FRUITS: Egg-shaped capsules

LOOKALIKES: Water Figwort (*Scrophularia aquatica*) is taller, with winged stems, blunt leaves and spherical capsules, and grows in wet places.

Erect, hairless perennial, unpleasant-smelling when bruised, with short, thick, knobbly rhizome and square, unwinged stems; of woods, riverbanks and damp, shady places. It is widespread and common through most of Britain and Ireland, although rare in N Scotland. It is not a handsome plant, although a cream-variegated variant finds favour in cottage gardens. However, it has a long history of medicinal use, especially to treat skin complaints and to heal wounds.

FIGWORT AND FOXGLOVE FAMILY, SCROPHULARIACEAE

| J | F | M | A | M | J |
| J | A | S | O | N | D |

Toadflax
Linaria vulgaris

ID FACT FILE

HEIGHT:
30–80 cm

FLOWERS: Pale
yellow, 2–3 cm
long, with a dark
yellow central
patch, in long,
fairly loose
spikes; long,
slender spur

LEAVES: Very
narrow, spear-
shaped, entire,
pointed

FRUITS: Egg-
shaped capsules

LOOKALIKES: Other
snapdragons and
toadflaxes, most
of them garden
escapes, usually
have purple or
red flowers.

Erect perennial with slender, creeping rhizome; a widespread plant of grassland, hedge-banks and waste land; local in Ireland and N and W Scotland. The long spurs contain nectar; the flowers are pollinated by bees and bumblebees, some of which steal nectar by biting through the spur. This is one of our most handsome wild flowers and a feature of the late summer countryside. The first half of the scientific name denotes the similarity of the leaves to those of flax (Latin: *linum*).

J	F	M	A	M	J
J	A	S	O	N	D

Ivy-leaved Toadflax
Cymbalaria muralis

ID FACT FILE

HEIGHT:
10–60 cm

FLOWERS: Snap-dragon-like, lilac, violet or some-times white, with a yellow central spot, 9–15 mm long, solitary on long stalks

LEAVES: Long-stalked, 'ivy'-like, rounded or kidney-shaped, with 5–9 shallow lobes

FRUITS: Spherical capsules, on stiff, curved stalks

LOOKALIKES: None of the other toad-flaxes has the trailing habit of this plant.

Hairless, often purplish, trailing perennial, characteristically on walls, but sometimes on rocks, stony ground and even shingle beaches. Originally a native of Italy and adjacent parts of the Alps, it has escaped from gardens through-out W Europe and has been known in Britain since 1640. It now occurs almost throughout Britain and Ireland. The stalks of the capsules grow away from the light, curving downwards as the seeds ripen, and pushing the fruits into chinks and crannies, where the seedlings are better able to establish and grow.

J	F	M	A	M	J
J	A	S	O	N	D

Foxglove
Digitalis purpurea

ID FACT FILE

HEIGHT:
50–180 cm

FLOWERS: Broadly
tubular, pinkish-
purple or reddish-
pink, red-spotted
and hairy inside,
40–55 mm long,
in a long, dense
spike

LEAVES: Very
large, broadly
spear-shaped,
wrinkled, softly
hairy

FRUITS: Nearly
spherical, downy
capsules

LOOKALIKES:
Unlikely to be
confused with
any other wild
plant.

Conspicuous, erect, unbranched, greyish-
downy biennial or short-lived perennial; widespread and sometimes abundant in open
woods, scrub, heaths and banks on acid soils.
Where ground has been cleared or burned,
huge numbers may colour the whole landscape, as on new road-verges in Wales and
other parts of W Britain. This very poisonous
plant yields the drug digitalin, which has long
been used in medicine to slow the pace of the
heart-beat. The flowers are much visited by
bumblebees.

| J | F | M | A | M | J |
| J | A | S | O | N | D |

Brooklime
Veronica beccabunga

ID FACT FILE

HEIGHT:
20–50 cm

FLOWERS: Blue,
rarely pink,
almost flat,
4-lobed, 5–8 mm
across, in paired,
conical spikes
arising from each
leaf-pair;
2 stamens

LEAVES: In
opposite pairs,
oval to oblong,
short-stalked,
shallowly
toothed, blunt

FRUITS: Flattened
capsules,
splitting into
4 segments

LOOKALIKES: Blue
Water-speedwell
(*Veronica
anagallis-
aquatica*) has
broadly spear-
shaped, pointed
leaves and paler
blue flowers.

Somewhat fleshy, hairless perennial, with far-creeping, rooting, sprawling or ascending, hollow stems; common in wet places, ditches and streams, it is widespread in Britain and Ireland, although local in C and W Scotland. The leaves have a sharp taste and have been used as a salad. The blue flowers with a minute tube and only two stamens identify Brooklime as one of the speedwells; it belongs to a group of four similar perennial speedwells that occur in wet or marshy ground.

FIGWORT AND FOXGLOVE FAMILY, SCROPHULARIACEAE

Germander Speedwell, or Bird's-eye

Veronica chamaedrys

ID FACT FILE

HEIGHT:
10–30 cm

FLOWERS: Intense blue with a white eye, rarely lilac, c.1 cm across, in loose conical clusters of 10–20; 2 stamens

LEAVES: In opposite pairs, oval-triangular, toothed

FRUITS: Heart-shaped, hairy capsules, splitting into 2 segments

LOOKALIKES: Heath Speedwell (*Veronica officinalis*) has stems that are hairy all round and bluish-lilac flowers in dense clusters; it grows in dry grassland and on heaths.

Hairy perennial, with far-creeping, rooting, ascending stems with two opposite lines of white hairs; a common plant of wood margins and grassland, it occurs throughout Britain and Ireland, and is rare only in the Outer Hebrides and Orkney. This is one of the most elegant and attractive of all our wild flowers. Alas, the delicate petals fall easily, especially if the plant is picked: perhaps the origin of a superstition that harm will come to the eyes of the picker or of his or her mother.

FIGWORT AND FOXGLOVE FAMILY, SCROPHULARIACEAE

Common Field-speedwell
Veronica persica

| J | F | M | A | M | J |
| J | A | S | O | N | D |

All through mild winters.

ID FACT FILE

HEIGHT:
10–50 cm

FLOWERS: Blue, with darker bluish-violet veins, white markings and eye, 8–12 mm across, solitary on slender stalks; 2 stamens

LEAVES: In opposite pairs, short-stalked, oval or triangular, coarsely toothed

FRUITS: Sticky-hairy capsules with 2 divergent lobes

LOOKALIKES: Ivy-leaved Speedwell (*Veronica hederifolia*), with 3- to 5-lobed leaves, smaller, lilac or blue flowers, and stout, rounded fruits, grows in gardens and woods.

Prostrate or spreading, hairy annual of disturbed ground, especially good, cultivated soil. Although it arrived in Britain from SW Asia only in the early 19th century, it is now the commonest speedwell of cultivated land. These comprise a group of eight, all annuals, some of them now rare as a result of modern intensive agriculture. Common Field-speedwell, alone of the group, is one of our most successful weeds and can cause severe infestations of vegetable crops, allotments and gardens.

FIGWORT AND FOXGLOVE FAMILY, SCROPHULARIACEAE

Common Cow-wheat
Melampyrum pratense

J	F	M	A	M	J
J	A	S	O	N	D

ID FACT FILE

HEIGHT:
20–60 cm

FLOWERS: Yellow-ish-white or yellow, 10–18 mm long, in pairs, turned to the same side; leaf-like, slightly toothed bracts

LEAVES: In opposite pairs, rather narrow, oval or spear-shaped, pointed

FRUITS: 4-seeded capsules

LOOKALIKES: Yellow-rattle has toothed leaves and yellowish-green bracts; it occurs in grassland.

A slender, branched, mostly hairless annual of dry woods, scrub, heaths and grassland on acid soils. It is widespread in the British Isles, although local in E England and much of Ireland. A very variable species: plants on moors sometimes have pinkish-purple flowers and another striking variant has golden-yellow flowers. Like the eyebrights and some other members of the family, Common Cow-wheat is a semi-parasite, taking water and minerals from the roots of other plants.

FIGWORT AND FOXGLOVE FAMILY, SCROPHULARIACEAE

J	F	M	A	M	J
J	A	S	O	N	D

Greater Yellow Rattle
Rhinathus angustifolius

ID FACT FILE

HEIGHT
10–80 cm

FLOWERS
Yellow, with
2 short, violet
teeth, 15–25
mm, long, in
loose spikes;
calex inflated and
persistent in fruit;
conspicuous,
yellowish-green
bracts. Flowers
from June to
September

LEAVES
In opposite nar-
row pairs, nar-
row, oblong or
spear-shaped,
toothed, pointed

FRUITS
Capsules, each
within inflated
calyx; seeds flat,
winged

LOOKALIKES
Similar to Yellow
Rattle but can be
distinguished by
the bracts, which
are green

This is a rare species, known only in about ten places in Surrey, Worcestershire, Lincolnshire, Yorkshire and Angus. Pollinated by bees or self-pollinated, it is found in meadows, waste grounds, arable land and on sand dunes. The bracts are yellow-green, the corolla tube curves upwards and the teeth of the upper lip of the corolla are twice as long as it is broad. The name derives from the rattling of the ripe seeds inside the dried-out inflated calyx.

BROOMRAPE FAMILY, OROBANCHACEAE

Common Broomrape
Orobanche minor

ID FACT FILE

HEIGHT:
10–80 cm

FLOWERS: Corolla
5-lobed, yellow-
ish, tinged violet,
pink or purple,
10–15 mm long,
in a loose,
cylindrical spike

LEAVES: Scale-like

FRUITS: Cylindrical
capsules,
containing many
dust-like seeds

LOOKALIKES: The
commonest of
12 native
species of
broomrape. The
closely related
Toothwort
(*Lathraea
squamaria*), with
1-sided clusters
of whitish or pink
flowers, is a
rather scarce
plant, parasitic
on Hazel and
Elm.

Erect, fleshy, sticky-hairy, yellow, brownish,
pink or purplish perennial, superficially resembling an orchid. Often erratic in appearance, it
is widespread but local in dry grassland, less
often in gardens. This plant produces no green
chlorophyll and is parasitic on the roots of
numerous plants, especially members of the
clover and daisy families, including
garden plants. From these it extracts water,
minerals and sugars to sustain its own growth,
flowering and seed production.

PLANTAIN FAMILY, PLANTAGINACEAE

Greater Plantain
Plantago major

J	F	M	A	M	J
J	A	S	O	N	D

ID FACT FILE

Height: 5–40 cm, sometimes up to 60 cm

Flowers: Small, yellowish-green, in a dense, cylindrical spike; stamens lilac, fading to yellowish

Leaves: Oval or elliptical, usually with 5–9 parallel veins, stalked, generally tough and hairless, sometimes very large

Fruits: Egg-shaped capsules, opening by a lid; usually 8–12 seeds

Lookalikes: Ribwort Plantain (*Plantago lanceolata*) has spear-shaped leaves, narrowly egg-shaped flower-spikes and yellow stamens.

A tufted perennial, with rosettes of leathery leaves and fibrous stems that are well able to withstand trampling; abundant in grassy places and waysides, and on waste ground, especially by paths and gates. A variable species; plants from cultivated land and lake shores have pale green, hairy leaves with 3–5 veins and up to 30 seeds in each capsule. They are regarded as a distinct subspecies. The leaves were once used to dress wounds, and the plant has healing and soothing properties.

HONEYSUCKLE FAMILY, CAPRIFOLIACEAE

J	F	M	A	M	J
J	A	S	O	N	D

Honeysuckle
Lonicera periclymenum

ID FACT FILE

HEIGHT: 2–6 m

FLOWERS: Tubular, 2-lipped, 3–5 cm long, yellowish-cream tinged lilac or reddish, fading to orange, richly scented, in clusters

LEAVES: In opposite pairs, oval, very short-stalked, downy, slightly paler beneath

FRUITS: Clusters of shiny red berries

LOOKALIKES: Other honeysuckles sometimes escape from gardens, especially Perfoliate Honeysuckle (*Lonicera caprifolium*), which has fused pairs of leaves.

A woody climber, twining (clockwise) in hedges and amongst the branches of trees and shrubs, or near the ground in coastal heathland, very conspicuous when in flower. A familiar and much-loved wild flower that is renowned for its scent, especially at night, when the flowers are visited by moths. It is the food plant of the caterpillars of the White Admiral butterfly. The new leaves are one of the first signs of green in woodlands during late winter. The berries are poisonous.

VALERIAN FAMILY, VALERIANACEAE

J	F	M	A	M	J
J	A	S	O	N	D

Common Valerian
Valeriana officinalis

ID FACT FILE

HEIGHT:
50–150 cm

FLOWERS: Pale
pink or white,
5 mm across,
tubular with a
pouch-like spur,
scented, densely
grouped in a flat
head; 3 stamens

LEAVES: In oppo-
site pairs, com-
pound, or very
deeply lobed, the
margins toothed

FRUITS: 1-seeded,
2–5 mm, with a
crown of hairs

LOOKALIKES: Red
Valerian has a
longer-spurred
corolla that is
usually red; Marsh
Valerian (*Valeriana
dioica*), a smaller
plant with less
divided leaves,
occurs in marshes
northwards to
C Scotland.

Robust, erect, rather hairy perennial of damp
and dry grassland, scrub and open woods,
widespread throughout Britain and Ireland.
This and other species of valerian have a curious, rather unpleasant, but characteristic
smell; the root is said to be irresistible to cats.
An extract from the plant has been used medicinally as a sedative – hence the scientific
name (Latin: *valere*, to heal) – and 'Valerian
drops' feature as a poison in old-fashioned
crime stories.

SCABIOUS FAMILY, DIPSACACEAE

J	F	M	A	M	J
J	A	S	O	N	D

Teasel
Dipsacus fullonum

ID FACT FILE

HEIGHT:
50–200 cm,
sometimes up to
300 cm

FLOWERS: Violet,
in an egg-shaped
head 3–9 cm
long; basket of
8–12 bracts,
narrow, upward-
curved, spiny, as
long as the
flowerhead

LEAVES: In oppo-
site, fused pairs,
spear-shaped,
prickly beneath

FRUITS: 1-seeded,
massed in char-
acteristic heads

LOOKALIKES: A
distinctive plant;
two other teasels
are smaller and
much rarer.

A stately, robust, hairless, erect biennial of streamsides, damp, grassy places, waysides and waste ground. It is widespread in S Britain, extending northwards to Fife, but rare in the north and in Ireland. The cups formed by the bases of the fused leaves fill with rain and dew, drowning many insects. A distinct subspecies with spreading bracts, Fuller's Teasel, has long been used to raise the nap or pile of woollen cloth. A crop is still grown near Taunton in Somerset for this purpose.

SCABIOUS FAMILY, DIPSACACEAE

| J | F | M | A | M | J |
| J | A | S | O | N | D |

Devil's-bit Scabious
Succisa pratensis

ID FACT FILE

HEIGHT:
20–100 cm

FLOWERS: Dark
bluish-purple,
rarely pink or
white; corolla
4-lobed, the
outer lobes
larger than the
inner, in a long-
stalked, domed
head 18–25 mm
across

LEAVES: Basal
leaves in a
rosette; stem
leaves in oppo-
site pairs, ellipti-
cal, the upper
ones narrower

FRUITS: 1-seeded,
c.5 mm

LOOKALIKES: Field
Scabious and
Small Scabious
have lilac flowers
in heads more
than 25 mm
across.

Erect perennial of damp grassland and
marshes. It can be very abundant and ubiquitous in grassland and grassy coastal heathland,
as over large areas of W Ireland. The short,
thick rhizome has an abruptly cut-off end –
bitten off by the devil! This is the food plant of
the caterpillar of the rare, declining Marsh
Fritillary butterfly. The word scabious derives
from the former herbal use of this and related
plants to cure scabies and other unpleasant
skin complaints.

SCABIOUS FAMILY, DIPSACEAE

J	F	M	A	M	J
J	A	S	O	N	D

Field Scabious

Knautia arvensis

ID FACT FILE

HEIGHT:
30–100 cm

FLOWERS: Lilac, rarely white; corolla 4-lobed, the outer lobes larger than the inner, in a long-stalked, flat head 25–40 mm across

LEAVES: In opposite pairs, the lower entire, the upper deeply lobed

FRUITS: 1-seeded, 5–6 mm long

LOOKALIKES: Small Scabious (*Scabiosa columbaria*) is shorter, with smaller flower-heads and 5-lobed corollas; it occurs on grassland on lime-rich soils.

Erect, hairy biennial or perennial of grassland, dry banks and road-verges; formerly on cultivated land. It is widespread, although rare in W and N Scotland and much of W Ireland. The word Scabious derives from the use of this group of plants to cure scabies and other skin complaints; today they are prized more for ornament, both as wild flowers and in gardens. Britain's native scabious species are all important food plants for the caterpillar of the Chalkhill Blue butterfly.

BELLFLOWER FAMILY, CAMPANULACEAE

J	F	M	A	M	J
J	A	S	O	N	D

Harebell

Campanula rotundifolia

ID FACT FILE

HEIGHT:
10–50 cm

FLOWERS: Violet-blue, bell-shaped, 12–20 mm long, nodding, in open, loose, branched clusters; very narrow, pointed calyx-teeth

LEAVES: Lower ones heart-shaped, rounded, toothed, stalked; upper spear-shaped, slightly toothed

FRUITS: Broadly conical capsules, with many tiny seeds, nodding when ripe

LOOKALIKES: Other bellflowers are more robust and less elegant and delicate.

Slender, erect or ascending, little-branched, hairless, creeping perennial of dry grassland, hedge-banks, heaths, rocky ground and sand-dunes. It exudes a white, milky juice when cut. This plant is the Bluebell of Scotland, where the English Bluebell is known as Wild Hyacinth. A variable species that needs more study; the rather fine-looking plants that occur on and near western coasts have fewer, slightly larger flowers and are regarded by some botanists as a separate species.

J	F	M	A	M	J
J	A	S	O	N	D

Hemp Agrimony
Eupatorium cannabinum

ID FACT FILE

HEIGHT:
30–180 cm

FLOWERS: Pink or reddish-lilac, 5–6, in heads 3–5 mm across, without ray florets, in branched, flat-topped clusters

LEAVES: In opposite pairs, 3- to 5-lobed, the lobes spear-shaped, toothed, the central ones longer

FRUITS: Heads of 1-seeded fruits, each with a 'parachute' of white hairs

LOOKALIKES: The plant bears a superficial resemblance to Common Valerian, which has compound or deeply lobed leaves.

Erect, robust, leafy, downy perennial, with reddish stems, forming clumps in wet woods, marshes, hedgerows, damp, grassy places and scrub, sometimes on shingle beaches or amongst limestone rocks. A widespread plant, but mostly coastal in Scotland; local in Ireland, where a dwarf variant occurs in the Burren of Co. Clare. The second half of the scientific name reflects the similarity of the leaves to those of Hemp (*Cannabis sativa*), related to Hop.

DAISY AND DANDELION FAMILY, COMPOSITAE

J	F	M	A	M	J
J	A	S	O	N	D

Golden-rod
Solidago virgaurea

ID FACT FILE

HEIGHT:
20–180 cm

FLOWERS: In heads 3–5 mm across, without ray florets, massed in branched spikes

LEAVES: Dark green, somewhat leathery; lower leaves in a loose rosette, spoon-shaped; stem leaves spear-shaped, narrower

FRUITS: Heads of 1-seeded fruits, each with a crown of brownish hairs

LOOKALIKES: Garden Golden-rod (*Solidago altissima*), which is taller, with many tiny heads of flowers in dense, 1-sided clusters, often escapes.

Erect, little-branched, somewhat downy perennial of dry, open woods, heaths, hedgebanks and rocky ground, especially on well-drained, acid soils. It is widespread, but local in both the Irish and English midlands and in East Anglia. An infusion of this plant has been used for its healing properties. A very variable species: for example plants from the Burren of Co. Clare are dwarf and flower in June–July, whereas most plants are tall and flower in August–September.

DAISY AND DANDELION FAMILY, COMPOSITAE

Daisy
Bellis perennis

ID FACT FILE

HEIGHT: 5–20 cm

FLOWERS: Solitary heads 1–3 cm across, the disc florets yellow, the ray florets white, reddish or purplish below

LEAVES: All basal, spoon-shaped, stalked, bluntly toothed, slightly fleshy and leathery

FRUITS: Head of 1-seeded fruits, all without hairs

LOOKALIKES: Ox-eye Daisy, with which it often grows, is a very much larger and more robust plant than even cultivated daisies.

A tufted, downy perennial with erect stems. This familiar and almost ubiquitous wild flower is most closely associated with closely mown grass: even a small, urban lawn will have a few Daisies. The plant's native habitat of old, short grassland is now much reduced, although fine stands of Daisies can still be seen on seaside banks, or grazed hedge-banks inland, and in churchyards. Larger variants are grown in gardens. The flowers close in the evening and on dull or wet days.

DAISY AND DANDELION FAMILY, COMPOSITAE

J	F	M	A	M	J
J	A	S	O	N	D

Sea Aster
Aster tripolium

ID FACT FILE

HEIGHT:
20–80 cm, sometimes up to 150 cm

FLOWERS: Heads 10–25 mm across, the disc florets yellow, ray florets 10–30, mauve or lilac (or absent), in loose clusters

LEAVES: Spear-shaped, usually untoothed

FRUITS: Heads of 1-seeded fruits, each with a 'parachute' of whitish hairs

LOOKALIKES: Michaelmas Daisy (*Aster novae-belgii*) frequently escapes from gardens; it is taller, not fleshy, and forms clumps on waste ground and railway embankments.

Erect, fleshy annual, biennial or perennial; an abundant plant of saltmarshes all around the coast of Britain and Ireland, together with the banks of tidal rivers; also on sea-cliffs and rocks in the west, and in a few saline marshes in the W Midlands of England. The flowers of many plants in some parts of England – for example the Thames Estuary – lack ray florets. This plant was grown in gardens before the Michaelmas Daisy was introduced from N America in the 17th century.

DAISY AND DANDELION FAMILY, COMPOSITAE

| J | F | M | A | M | J |
| J | A | S | O | N | D |

Canadian Fleabane
Conyza canadensis

ID FACT FILE

HEIGHT:
30–180 cm

FLOWERS: Heads
3–5 mm across,
with white or
pinkish disc and
ray florets,
numerous in
long, branched
loose clusters

LEAVES: Narrow,
spear-shaped,
untoothed or
finely toothed

FRUITS: Heads of
1-seeded fruits,
each with a
'parachute' of
hairs

LOOKALIKES: The
cudweeds are all
much smaller,
white-woolly
annuals; Marsh
Cudweed has
flowers in flat-
topped clusters.

Erect, branched, leafy, pale green, hairy
annual of waste ground, derelict land, fallow
fields and sand-dunes. An early arrival from N
America in the 17th century, it is today especially common in SC and SE England, and has
long been a feature of waste ground in and
around London. In the 1980s it was reported
in Ireland, from Dublin. It may spread on
farmland set aside under EU rules, and further
expansion of its range is likely should global
warming modify our climate.

J	F	M	A	M	J
J	A	S	O	N	D

Trifid Bur-marigold
Bidens tripartita

ID FACT FILE

HEIGHT:
20–60 cm

FLOWERS: Heads
10–25 mm
across, disc
florets yellow, ray
florets usually
absent, in
branched clusters
with 5–8 leaf-like
bracts

LEAVES: In oppo-
site pairs, 3-lobed
(rarely 5-lobed),
the lobes spear-
shaped, coarsely
toothed

FRUITS: Heads of
flattened, wedge-
shaped, 1-seed-
ed fruits, each
with down-turned
marginal hairs
and 3–4 barbed
bristles

LOOKALIKES: The
Nodding Bur-
marigold (*Bidens
cernua*) has
spear-shaped
leaves and nod-
ding flowerheads.

A rather dowdy, erect, branched, often hairy
annual, with purplish, winged stems, that
occurs on the margins or dried mud of lakes,
ponds and rivers, ditches and in damp waste
places. Often growing in great, dense crowds,
it is generally widespread, but is rare north of
Cumbria and Galway Bay. The bristly fruits
adhere readily and firmly to clothing (especially woolly socks) and fur, which serves to disperse the seeds most effectively. Very rarely,
flowerheads have yellow ray florets.

DAISY AND DANDELION FAMILY, COMPOSITAE

J	F	M	A	M	J
J	A	S	O	N	D

Gallant Soldier
Galinsoga parviflora

ID FACT FILE

Height:
10–80 cm

Flowers: Heads of yellow florets 3–5 mm across, each with 5 (sometimes 4 or 6) small, white ray florets

Leaves: In opposite pairs, oval, with a few large, marginal teeth

Fruits: Tiny, oval, flattened, black, with minute hairs and a tuft of scales at one end

Lookalikes: Hairy Gallant Soldier (*Galinsoga quadriradiata*), another, but less common, introduced weed from S America, has hairy stems.

Almost hairless, branched annual of cultivated land; sometimes abundant in gardens, allotments, nursery beds and vegetable crops, especially in SE England. Introduced from Peru, it escaped from Kew Gardens during the 1860s and spread through much of Britain; during the 1980s it was reported from Ireland. It is an interesting example of a plant that acquired a popular name within a few years of its arrival here. Tiny bristles on the fruit, which readily adhere to clothing or fur, aid seed dispersal.

DAISY AND DANDELION FAMILY, COMPOSITAE

J	F	M	A	M	J
J	A	S	O	N	D

Pineapple Weed
Matricaria discoidea

ID FACT FILE

HEIGHT: 5–40 cm

FLOWERS: Almost spherical heads, 5–9 mm in diameter, of greenish-yellow disc florets; no ray florets

LEAVES: Compound, feathery, with numerous narrow segments

FRUITS: Heads of 1-seeded fruits, without hairs

LOOKALIKES: A distinctive plant. Scentless Mayweed and other mayweeds have conspicuous white ray florets, giving them a daisy-like appearance.

Erect, stiffly branched, aromatic annual, of pathsides, especially by field gates and other trampled places, waste ground and cultivated land. The whole plant smells strongly of pineapple (some say apple) when bruised. So widely is this undistinguished-looking but successful weed now distributed worldwide that its precise origin (probably western USA) is unknown. First recorded in Britain in 1871, and Ireland in 1894, it spread over much of these islands in the first quarter of this century.

DAISY AND DANDELION FAMILY, COMPOSITAE

J	F	M	A	M	J
J	A	S	O	N	D

Yarrow
Achillea millefolium

ID FACT FILE

HEIGHT:
10–100 cm

FLOWERS: Heads
3–6 mm across,
with white or
cream disc florets
and 5 white or
pinkish-purple ray
florets, in flat-
topped clusters

LEAVES: Fern-like
and feathery,
with numerous
narrow segments

FRUITS: Heads of
nut-like, 1-seeded
fruits, without
hairs

LOOKALIKES:
Sneezewort
(*Achillea
ptarmica*), of
damp heaths and
marshes, has
undivided,
toothed leaves,
and clusters of
flowerheads
12–18 mm
across, with
greenish disc and
white ray florets.

Erect, tough-stemmed, hairy, aromatic perennial, forming clumps and patches; a common
plant of dry grassland, hedge-banks and waste
places, a prominent feature of village greens,
roadsides and untended lawns in late summer.
This plant is drought-tolerant and during dry
spells can be seen, not only green but covered
with flowers, on brown lawns and dry road-
verges. A variable species in size and flower
colour: several colour variants are favourite
plants of the cottage garden.

DAISY AND DANDELION FAMILY, COMPOSITAE

J	F	M	A	M	J
J	A	S	O	N	D

Tansy
Tanacetum vulgare

ID FACT FILE

HEIGHT:
50–150 cm

FLOWERS: Heads
7–12 mm
across, golden
yellow, without
ray florets,
10–70 in flat-
topped clusters

LEAVES: Fern-like,
oblong, deeply
lobed, toothed,
dark green

FRUITS: Heads of
ribbed, 1-seeded
fruits, without
hairs

LOOKALIKES: Fever-
few or Bachelor's
Buttons (*Tanace-
tum parthenium*)
is smaller, with
yellowish-green
leaves, and
flowerheads with
white ray florets;
it rarely occurs
far from houses.

Erect, leafy, sweetly aromatic perennial, forming clumps, the stems branched in the upper
part; it is a plant of waste places, roadsides,
hedgerows and riverbanks, often around
houses or ruined buildings. It has long been
used medicinally and is an ingredient of the
Tansy puddings formerly eaten at Easter, and
of drisheen, a spicy black pudding from SW
Ireland. Like other aromatic members of the
daisy and dandelion family, the fresh leaves
discourage flies and other insects.

Ox-eye, or Moon, Daisy
Leucanthemum vulgare

ID FACT FILE

HEIGHT:
20–100 cm

FLOWERS: Heads
25–50 mm
across, with yel-
low disc florets
and white ray
florets; bracts
with brown or
black margin

LEAVES: Oval to
spoon-shaped,
toothed, dark
green; upper
ones oblong,
clasping the
stem

FRUITS: Heads of
1-seeded fruits,
without hairs

LOOKALIKES: A
much larger plant
than Daisy,
which often
occurs with it in
grassland.

Erect, usually unbranched, short-lived perennial; a characteristic plant of old meadows,
churchyards and coastal grassland and sand-
dunes. A variable species: for example, dwarf
plants occur on sea-cliffs and coastal heaths. In
recent years it has become very common and
conspicuous on newly landscaped motorway
verges. Robust, branched plants with large
leaves and flowers on road-verges derive from
commercial wild flower seed, and are probably
of garden origin.

DAISY AND DANDELION FAMILY, COMPOSITAE

J	F	M	A	M	J
J	A	S	O	N	D

Mugwort
Artemisia vulgaris

ID FACT FILE

HEIGHT:
50–180 cm

FLOWERS: Tiny, in reddish-brown, egg-shaped heads, 2–3 cm across, without ray florets, grouped in loose clusters

LEAVES: Much-lobed, the lobes spear-shaped or themselves deeply lobed, hair-less above, white-hairy beneath

FRUITS: Heads of nut-like, 1-seeded fruits, without hairs

LOOKALIKES: Wormwood (*Artemisia absinthium*) is aromatic, with silky-hairy, silvery leaves, and also occurs in waste places near houses or ruins.

Tough, erect, untidy perennial of dry waysides, waste ground and derelict land, almost always near buildings or roads. In late summer this is the most typical plant flowering on dusty, urban and suburban roadsides, demolition sites, neglected pavements and waste ground. The dried leaves have served as a substitute for tobacco. Like St John's Wort, it was used to ward off evil spirits as part of the St John's Eve pagan festivities at the Summer Solstice.

DAISY AND DANDELION FAMILY, COMPOSITAE

| J | F | M | A | M | J |
| J | A | S | O | N | D |

Colt's-foot
Tussilago farfara

ID FACT FILE

HEIGHT:
10–30 cm

FLOWERS: Heads on stout, scaly stems, yellow; disc florets few, the ray florets orange beneath, in heads 20–35 mm across

LEAVES: Appearing after flowers, all basal, heart-shaped, up to 25 cm across, toothed, cobwebby beneath (also above at first)

FRUITS: Heads of 1-seeded fruits, each with a 'parachute' of long hairs

LOOKALIKES: Dandelion has ray florets only, and the flowers are produced on leafless stems at the same time as leaves.

Erect perennial, with thick, creeping rhizome, forming large patches on the bare ground of waste places, road-verges, banks of rivers and streams, and low sea-cliffs. Often found on clay soils, it is an early colonist of clay banks of rivers and at the seaside. It is one of the plants that occupies ground left bare by retreating Alpine glaciers. Colt's-foot has long held a reputation as a cure for coughs and illnesses of the chest. The flowers are one of the first indicators of spring.

DAISY AND DANDELION FAMILY, COMPOSITAE

J	F	M	A	M	J
J	A	S	O	N	D

Butterbur
Petasites hybridus

ID FACT FILE

HEIGHT:
10–30 cm, female plants elongating to 60 cm

FLOWERS: Pale lilac or yellowish, unscented, in large heads

LEAVES: Appearing after flowers, all basal, rhubarb-like, up to 1 m across, toothed, cobwebby beneath (also above at first)

FRUITS: Heads of 1-seeded fruits, each with a 'parachute' of hairs

LOOKALIKES: Winter Heliotrope (*Petasites fragrans*) has richly scented flowers in looser clusters, produced from November to March, with the leaves; it is a garden escape.

Erect perennial, with a creeping rhizome, forming large, conspicuous patches; rather local on the banks of streams and rivers, in damp woods and on road-verges. It prefers sandy or well-drained soils. The stout clusters of flowers are an impressive sight in early spring; later the immense leaves form great thickets. Most plants are male. The female plant, with flowerheads that elongate considerably in fruit, is very local in occurrence, mainly in parts of N England.

DAISY AND DANDELION FAMILY, COMPOSITAE

J	F	M	A	M	J
J	A	S	O	N	D

All through mild winters.

ID FACT FILE

HEIGHT: 5–40 cm

FLOWERS: Yellow, in shaving-brush-like heads 4–5 mm across, usually without ray florets, in loose clusters

LEAVES: Coarsely and bluntly lobed and irregularly toothed

FRUITS: Heads of 1-seeded fruits, each with 'para-chute' of hairs

LOOKALIKES: Other species of groundsel or ragwort have conspicuous ray florets or sticky-hairy bracts.

Groundsel
Senecio vulgaris

Ubiquitous, erect, branched, rather hairy annual of waste places, walls, paths and pavements, cultivated ground, sand-dunes and shingle beaches. Groundsel is one of the first flowers of late winter and early spring, along with Chickweed, Shepherd's Purse and Red Dead-nettle. Occasional plants with short ray florets may derive from crossing with Oxford Ragwort. The first half of the scientific name refers to the white fruiting heads (Latin: *senex*, old man).

Ragwort
Senecio jacobaea

| J | F | M | A | M | J |
| J | A | S | O | N | D |

ID FACT FILE

HEIGHT:
30–150 cm

FLOWERS: Bright yellow, in heads 15–25 mm across, with ray florets present, in often dense, flat-topped clusters

LEAVES: Lyre-shaped or deeply lobed, irregularly toothed, usually hairy beneath; upper leaves clasping the stem

FRUITS: Heads of 1-seeded fruits, each with a 'parachute' of hairs

LOOKALIKES: Oxford Ragwort has lobed, less complexly dissected, toothed leaves, and laxer clusters of flowerheads.

An erect, leafy biennial or short-lived perennial of waysides, waste ground, pastures, sand-dunes and shingle beaches. The leaves give off an unpleasant smell when bruised, hence local names such as 'Stinking Willie'. It is a poisonous plant that can be a great menace to livestock, especially horses. The caterpillars of the Cinnabar Moth feed on Ragwort, accumulating its poisonous chemicals to protect themselves against birds, hence their yellow- and black-striped warning colours.

DAISY AND DANDELION FAMILY, COMPOSITAE

| J | F | M | A | M | J |
| J | A | S | O | N | D |

Lesser Burdock
Arctium minus

ID FACT FILE

HEIGHT:
50–150 cm

FLOWERS: Purple, in spherical heads, 15–20 mm across, in long, loose clusters; bracts dense, hooked

LEAVES: Heart-shaped, up to 50 cm long, cottony beneath, with long, hollow stalks; upper leaves smaller, narrower

FRUITS: Egg-shaped 'bur' heads 20–35 mm across, enclosed by involucre of stiff, hooked hairs; each head is dispersed as a unit

LOOKALIKES: Great Burdock (*Arctium lappa*), has fewer flowerheads, 30–45 mm across.

A robust, leafy, downy biennial, with many, branched stems, common in dry woods, on roadsides and in waste places. The hooked burs adhere readily to clothing or fur, dispersing the seeds. At an annual fair in August, the Burry Man, his costume densely studded with these burs, parades in strict silence around South Queensferry, West Lothian. The plant is such a feature of woods and waysides that it sometimes appeared as a detail in the paintings and sketches of John Constable.

DAISY AND DANDELION FAMILY, COMPOSITAE

Common, or Spear, Thistle

Cirsium vulgare

J	F	M	A	M	J
J	A	S	O	N	D

ID FACT FILE

HEIGHT:
30–180 cm

FLOWERS: Purple, scented, in egg-shaped heads 3–5 cm across, 1–3 in branched clusters; bracts spine-tipped

LEAVES: Spear-shaped, deeply lobed, with coarse, irregular lobes, each with a spine

FRUITS: Heads of 1-seeded fruits, each with a 'para-chute' of soft, feathery hairs (thistle-down)

LOOKALIKES: Creeping Thistle (*Cirsium arvense*) is a perennial forming great patches; flower-heads, mauve, 15–20 mm across, with purple bracts, in loose clusters.

Erect, leafy, ferociously spiny, hairy annual or biennial of pastures, roadsides and waste ground. This is probably the true Scots Thistle; the larger thistle of that name is a rare plant in Scotland. It is said that thistles impeded a surprise night attack by the Danes during the Battle of Largs by pricking the attackers, whose consequent cries of pain alerted the defenders. Like other thistles, the flowers are attractive to bumblebees. The fruits are often eaten by beetle larvae.

DAISY AND DANDELION FAMILY, COMPOSITAE

Creeping Thistle
Cichorium arvense

ID FACT FILE

HEIGHT:
60–120 cm

FLOWERS: Pale
violet, in heads
1.5–2.5 cm
across, bracts
purplish

LEAVES: Grey-
green and hair-
less with wavy,
toothed edges;
lower leaves
stalked

FRUITS: 3mm,
brown, no spots

LOOKALIKES: Brook
Thistle *C. rivulare*
is hairier with
larger, rounder
flowerheads

The most common species of thistle, this is a
creeping perennial, erect, branched and spineless with an unwinged stem. The fragrant flowerheads are pale violet and arranged in great
flat clusters. Its creeping roots enable it to
form large colonies in fields, cultivated and
waste ground, footpaths and clearings all over
the British Isles. All types of thistles have been
eaten as food throughout history, with stems
and leaves used in salads.

DAISY AND DANDELION FAMILY, COMPOSITAE

Chicory
Cichorium intybus

ID FACT FILE

HEIGHT:
30–120 cm

FLOWERS: Pale
blue, in heads
3–5 cm across,
1–3 in axils of
upper leaves

LEAVES: Basal
leaves in a
rosette, deeply
and coarsely
lobed; upper
leaves spear-
shaped, pointed

FRUITS: Heads of
nut-like, 1-seed-
ed fruits, without
hairs

LOOKALIKES: Blue
Sowthistle
(*Cicerbita
macrophylla*) is
erect, up to 2 m
tall, the flowers in
branched, flat-
topped clusters;
introduced, mainly
on roadsides.

Erect or spreading, untidy perennial, with
stiffly branched, zigzag, grooved stems, of
waste places, dry waysides and field margins. It
is commonest on lime-rich soils. The flowers
look their best in the morning, fading after
midday. Chicory is widely cultivated, especially
in gardens, where it has long been a valued
winter salad. It also has medicinal (mainly
diuretic and laxative) properties, and the roast-
ed roots are used as a substitute for coffee,
especially in C Europe.

| J | F | M | A | M | J |
| J | A | S | O | N | D |

All through mild winters.

Dandelion

Taraxacum officinale

ID FACT FILE

Height: 5–40 cm

Flowers: Yellow, 25–40 mm across, solitary on long, hollow stems, the outer ray florets often greenish, brownish or red-dish beneath

Leaves: In a basal rosette, deeply lobed or toothed

Fruits: Heads of 1-seeded fruits, each with a 'parachute' of feathery hairs

Lookalikes: The smooth, hollow stems, large, solitary flowers, and down-turned bracts, distin-guish this vari-able species.

Familiar, tufted, very sparsely hairy perennial, of grasslands, road-verges, sand-dunes, wet places, open and disturbed ground, notably as a garden weed of lawns and flower-beds. In recent years, being slightly salt-tolerant, it has spread along road-verges on a massive scale; in April they can be a magnificent sight. The root makes an agreeable substitute for coffee, and dandelion latex provided the former Soviet Union with rubber during World War II. A very variable plant that botanists divide into hundreds of similar 'microspecies'.

DAISY AND DANDELION FAMILY, COMPOSITAE

| J | F | M | A | M | J |
| J | A | S | O | N | D |

Autumn Hawkbit
Leontodon autumnalis

ID FACT FILE

Height:
10–60 cm

Flowers: Yellow,
12–35 mm
across, solitary,
erect in bud, the
outer florets
often reddish-
streaked beneath

Leaves: All in a
basal rosette,
very deeply
lobed, shiny

Fruits: Heads of
1-seeded fruits,
each with a
'parachute' of
feathery hairs

Lookalikes:
Lesser Hawkbit
(*Leontodon
taraxacoides*), of
more lime-rich
soils, has shal-
lowly lobed
leaves and
flowerheads
12–20 mm
across, nodding
in bud, the outer
florets greyish-
violet beneath.

Erect, somewhat branched, sparsely hairy
perennial, widespread and frequently abundant in dry and damp grassland, waysides and
open stony ground such as lake-shores, especially on more acid soils. This is the most conspicuous of the yellow dandelion-like wild
flowers that appear in late summer and early
autumn. The hawkbits and their relatives are
very difficult to tell apart, requiring careful
examination of leaf-hairs, and other small
features, with a lens.

DAISY AND DANDELION FAMILY, COMPOSITAE

| J | F | M | A | M | J |
| J | A | S | O | N | D |

Goat's-beard
Tragopogon pratensis

ID FACT FILE

HEIGHT:
30–70 cm

FLOWERS: Yellow, in solitary, long-stalked heads, 3–5 cm across; bracts 8, up to 3 cm long, pointed

LEAVES: Grass-like, rather grey-ish-green and fleshy, pointed, closely sheathing the stem

FRUITS: Conspicuous heads of 1-seeded fruits, each with a large 'parachute' of stiff, feathery hairs

LOOKALIKES: This plant has an unmistakable combination of grass-like leaves and yellow, dandelion-like flowers.

Erect, annual, biennial or short-lived perennial, arising from a narrow, parsnip-like root, of dry grassland, road-verges and waste places. The flowers close after midday, giving the plant its local name of 'Jack- (or John-)-go-to-bed-at-noon'. The closely related, similar but purple-flowered Salsify (*Tragopogon porrifolius*) is sometimes grown as a root vegetable. The large fruits demonstrate clearly the features of the 'dandelion-type' seed-fruit unit and dispersal mechanism.

DAISY AND DANDELION FAMILY, COMPOSITAE

Corn, or Perennial, Sow-thistle

Sonchus arvensis

ID FACT FILE

HEIGHT:
50–150 cm

FLOWERS: Golden yellow, in heads 3–5 cm across, in loose clusters; bracts and flower-stems with yellowish hairs, very rarely hairless

LEAVES: Oblong to spear-shaped, deeply lobed, spiny-toothed; upper leaves clasping stem

FRUITS: Heads of 1-seeded fruits, each with a 'parachute' of hairs

LOOKALIKES: Smooth Sow-thistle and Prickly Sow-thistle (*Sonchus asper*) are annual, with paler flowerheads 20–25 mm across.

Erect, leafy perennial, with a network of far-creeping rhizomes and more slender roots; forming sometimes quite extensive patches on cultivated land, field margins, banks of streams and rivers, sand-dunes and shingle beaches. Even quite small fragments of root are able to grow into new plants, making this a noxious weed if it is not controlled. This is one of the handsomest and most conspicuous wild flowers of the late summer and early autumn countryside.

DAISY AND DANDELION FAMILY, COMPOSITAE

Smooth Sow-thistle

Sonchus oleraceus

ID FACT FILE

HEIGHT:
30–120 cm

FLOWERS: Rather
pale yellow, the
ray florets pur-
plish beneath, in
numerous heads
20–25 mm
across; bracts
mostly hairless

LEAVES: Variable in
shape, deeply
lobed, with a large
terminal lobe,
spiny-toothed;
upper leaves
clasping stem

FRUITS: Heads of
1-seeded fruits,
each with a 'para-
chute' of hairs

LOOKALIKES: Prick-
ly Sow-thistle
(*Sonchus asper*)
is often taller and
less branched,
with more clus-
tered flower-
heads, prickly
leaves and
sticky-hairy
bracts.

Erect, branched, often reddish-tinged annual; occurs mostly on cultivated ground where it can be a troublesome weed, although it will grow almost anywhere, even on the tops of walls or in the crevices of tarmac. The French feed this plant to edible snails, whilst in Greece and elsewhere it is eaten as a winter salad, as it was once (like many dandelion relatives) in England. The first half of the scientific name derives from *Sonchos*, the ancient Greek name for this plant.

DAISY AND DANDELION FAMILY, COMPOSITAE

Nipplewort
Lapsana communis

J	F	M	A	M	J
J	A	S	O	N	D

ID FACT FILE

HEIGHT:
30–120 cm

FLOWERS: Pale
yellow, in
numerous heads
15–20 mm
across, in loose,
branched
clusters

LEAVES: Oval,
deeply lobed with
large terminal
lobe, toothed;
upper stalkless,
spear-shaped,
untoothed

FRUITS: Narrowly
egg-shaped
heads of nut-like,
1-seeded fruits,
without hairs,
partly enclosed
by the bracts

LOOKALIKES: The
fruits, without
hairs, are unique
among yellow
dandelion-like
flowers.

Erect, branched, leafy annual or biennial, hairy
in lower part: a common plant of waysides,
hedgerows, wood margins, shady gardens and
cultivated land. It can sometimes be a troublesome weed. It grows best on heavier clay soils.
The young leaves were once used as a salad,
and slightly larger, more long-lived plants are
found here and there, which may be relics of
former cultivation. The common name derives
less from any healing properties than from a
reference to the shape of the buds.

DAISY AND DANDELION FAMILY, COMPOSITAE

Mouse-ear Hawkweed

Pilosella vulgaris

ID FACT FILE

HEIGHT: 5–20 cm, sometimes up to 35 cm

FLOWERS: Pale yellow, the florets usually red-striped beneath, 18–25 mm across, solitary, on long, leafless stems

LEAVES: Basal ones in a rosette, oblong to spear-shaped, blunt, with long, white hairs above, white-downy beneath

FRUITS: Heads of 1-seeded fruits, each with a 'parachute' of hairs

LOOKALIKES: Fox-and-Cubs or Grim-the-Collier (*Pilosella aurantiacum*), with blackish hairs and clustered orange flowerheads, is a garden escape.

Hairy perennial, with erect flowering stems and extensively creeping, leafy runners, forming often large patches in open ground or short turf of dry pastures, heaths, rocky ground, banks, walls and lawns. It is a characteristic plant of dry road and railway cuttings, banks on old lawns, and earth-filled stone walls, especially on sandy soils or near the sea. The hawkweeds are divided by botanists into many 'microspecies', but this plant is easily recognised by the pale yellow flowers and long runners.

WATER-PLANTAIN FAMILY, ALISMATACEAE

Arrowhead
Saggitaria sagittifolia

| J | F | M | A | M | J |
| J | A | S | O | N | D |

ID FACT FILE

HEIGHT
30–90 cm

FLOWERS
Flowers are white with a red base, 1.5–2 cm in diameter, 3 petals. Its flowering season is from June to August

LEAVES
Arrow-shaped leaves which grow out above the water surface in the flowering season

FRUITS
Flattish head of 1-seeded fruits in a ring

LOOKALIKES
Water plantain leaves are oval to lanceolate. The flowers are smaller than Arrowhead, open in the afternoon, have 3 green outer 3 white inner segments.

Commonly found throughout England, but rarely in Wales and the south-west and infrequently in Ireland, this hairless water plant grows to 1 m tall. Preferring fertile clays and loam soils, it grows in shallow, unpolluted water of ponds, ditches, streams, canals, dykes, and slow-flowing rivers. Flowering during July and August, it is a submerged floating or emergent aquatic perennial.

WATER-PLANTAIN FAMILY, ALISMATACEAE

J	F	M	A	M	J
J	A	S	O	N	D

Common Water-plantain
Alisma plantago-aquatica

ID FACT FILE

HEIGHT:
30–100 cm

FLOWERS: White or
pale lilac, up to
10 cm across, in
loose, long-
stalked, domed
clusters;
3 petals and
sepals

LEAVES: All basal
in a rosette, oval
to spear-shaped,
rounded or heart-
shaped at base

FRUITS: Flattish
head of 1-seeded
fruits in a ring

LOOKALIKES: Arrow-
head (*Sagittaria
sagittifolia*) has
arrow-shaped
leaves, larger
flowers and fruits
in a spherical
cluster; it occurs
in ponds, canals
and slow-moving
rivers.

Erect, hairless, aquatic perennial of lakes,
marshes, canals and ponds; it survives even in
the muddiest, scruffiest ponds and streams. It
is widespread in Britain and Ireland, although
rare in N Scotland. Individual flowers open
after midday and usually wither within 24
hours. This is the commonest of five native
water-plantains, which resemble the buttercups and water-buttercups, having similar
flowers (but only three, not five, petals), many
stamens and one-seeded fruits.

FLOWERING RUSH FAMILY, BUTOMACEAE

Flowering Rush
Butomus umbellatus

ID FACT FILE

Height:
50–150 cm

Flowers: Pink with darker veins, 25–30 mm across, many, in a domed cluster or umbel; 6 perianth-segments; carpels 6, stamens 9, both red

Leaves: All basal, 3-angled, narrow, rush-like

Fruits: Head of purple capsules, united in a ring

Lookalikes:
Common Water-plantain and Arrowhead have smaller, white flowers and broader leaves.

A striking, erect, hairless, aquatic perennial, with a creeping rhizome, of slow streams and rivers, reed-beds and ditches; widespread though local in England, but rare in Wales (except Anglesey), Scotland and most of Ireland. This is one of our handsomest aquatic plants, which can add grandeur to the shabbiest canal bank; hence the local name of 'Pride-of-the-Thames', commemorating the river where John Gerard, author of the famous 1597 *Herball*, knew it in the 16th century.

LILY AND BLUEBELL FAMILY, LILIACEAE

J	F	M	A	M	J
J	A	S	O	N	D

Solomon's-seal
Polygonatum multiflorum

Fairly common in south-west England, in scattered locations in the north of England, Solomon's-seal is a plant of the dry woodlands, found especially on chalk and limestone. Occasionally it occurs on acid soils. Pollinated by bumble-bees or self-pollinated, its native distribution is hidden by the regions to which it escapes from cultivation, sometimes reaching as far north as Scotland.

ID FACT FILE

HEIGHT
30–60 cm.

FLOWERS
Flowers are unscented in clusters of 2–5, drooping from leaf axils. Flowering season, June and July.

LEAVES
Leaves are alternate, ovate, to 15 cm long.

FRUITS
6–8 mm in diameter. A dark blue berry about 1 cm across.

LOOKALIKES
P. verticillatum, Whorled Solomon's-seal, also has an angled stem. Lanceolate to linear upper stem leaves, mostly in whorls of 3. Rare in Britain, found in Scotland only.

IRIS AND CROCUS FAMILY, IRIDACEAE

J	F	M	A	M	J
J	A	S	O	N	D

Yellow Flag
Iris pseudacorus

ID FACT FILE

HEIGHT:
50–150 cm

FLOWERS: 8–10 cm across, yellow, 4–12 in a cluster, usually 1–3 open at once, each with a sheathing bract; sepals broad, nodding

LEAVES: Evergreen, narrow, spear-shaped, sharp-edged, pointed

FRUITS: Cylindrical capsules 4–8 cm long, splitting into 3 segments; seeds brown

LOOKALIKES: Stinking Iris (*Iris foetidissima*), with violet or sometimes pale yellow flowers, narrower leaves and scarlet seeds, is a plant of woods, hedges and sand-dunes, mainly in S England.

Erect perennial with stout rhizomes and robust flowering stems, common in marshes, ditches and on the shores of rivers, canals, lakes, ponds and flooded gravel-pits; widespread and frequently abundant throughout Britain and Ireland. The rhizome yields a black dye. One of our stateliest and most distinctive wild flowers, this handsome plant has survived despite the destruction of so many wetland habitats. The flowers are much visited by bumblebees. The seeds float and are dispersed by water.

BUR-REED FAMILY, SPARGANIACEAE

J	F	M	A	M	J
J	A	S	O	N	D

Branched Bur-reed

Sparganium erectum

ID FACT FILE

HEIGHT:
30–150 cm

FLOWERS: Small, grouped in male and female, spherical heads on 3–8 branches; male clusters above, smaller, yellow

LEAVES: Narrow, strap-shaped, stiff; sometimes floating

FRUITS: Bur-like spherical clusters

LOOKALIKES: Unbranched Bur-reed (*Sparganium emersum*), less common, has more or less unbranched stems; it has erect or floating leaves.

Robust, rhizomatous, erect, hairless perennial, of marshes and shallow water and exposed mud in ditches, ponds and disused canals; occurs throughout Britain and Ireland, although local in many districts. There are three other native bur-reeds, usually with floating leaves in deeper water, all of which have the similar distinctive flowerheads; only two are at all widespread, the two rarer species being found mostly in peaty pools and lakes in W Scotland and W Ireland.

BULRUSH FAMILY, TYPHACEAE

J	F	M	A	M	J
J	A	S	O	N	D

Bulrush or Reedmace
Typha latifolia

ID FACT FILE

HEIGHT: 1–3 m

FLOWERS: Tiny, the female ones in cylindrical, sausage-like spikes 8–15 cm long; the male ones immediately above

LEAVES: Strap-shaped, 8–20 mm wide

FRUITS: Tiny, clustered in the dark brown, cylindrical spikes; seeds packed amongst soft, brownish hairs

LOOKALIKES: Lesser Bulrush or Reedmace (*Typha angustifolia*) has leaves c.4 mm wide and male and female flower-clusters slightly separated.

Robust, reed-like, aquatic perennial with creeping rhizome; forming dense stands in reed-swamps, lakes, ponds and slow-flowing streams and rivers. The spikes of fruit explode when ripe and dry, releasing huge numbers of seeds and the hairs (which represent reduced perianth-segments) that pack them within the 'mace'. This is the true Bulrush, as depicted in Victorian illustrations of the infant Moses, although botanists often apply this name to the Common Club-rush, which is related to the sedges.

ORCHID FAMILY, ORCHIDACEAE

Heath Spotted Orchid
Dactylorhiza maculata

ID FACT FILE

HEIGHT
15-60 cm

FLOWERS
Pale lilac. 3-lobed lip with a wavy edge and a single, small tooth

LEAVES
Broadly eliptic and pointed, the leaves cluster together at the base of the stem

FRUITS
Twisted cylindrical capsules containing dust-like seeds

LOOKALIKES
Common Spotted Orchid, the most common of native orchids. The lobes are deeper and fatter than those of the Heath-spotted Orchid.

A short to medium perennial, locally common in the British Isles. The Heath Spotted Orchid prefers heaths and moorland with acid, mineral and peat soils. More common than the Common Spotted Orchid in the north and west, favouring peaty roadsides, where it can be found in significant numbers. The Common Sotted Orchid can be found in open woods, scrub, fens and grassland, usually on chalky soil, it varies in colour from white to deep pink.

ORCHID FAMILY, ORCHIDACEAE

Broad-leaved Helleborine
Epipactis helleborine

J	F	M	A	M	J
J	A	S	O	N	D

ID FACT FILE

HEIGHT:
30–80cm

FLOWERS:
Variable; green-ish-yellow to purple-red.
Unscented.
Lip recurved and appears roun-ded.

LEAVES:
Short, almost rounded, alter-nate, spiralling up the stem.

LOOKALIKES:
Violet Helle-borine *E. purpu-rata* flowers later and has a purple sheen to leaves and stem.
Leaves are nar-row and flowers are greenish-white.

Purple-tinged perennial, uncommon but wide-spread in Britain. It is a plant of old woods and woodland margins, favouring chalk or base rich soils. However in Glasgow, colonies are found in much more varied habitats inclusing parks, cemeteries, golf-courses, gardens, coal spoil heaps and railway embankments. Indeed, there is even a hybrid species *E. youngiana*, of which *E. helleborine* may be one parent, which is endemic to northern Britain.

PINE FAMILY, PINACEAE

Evergreen
Up to 65 m

| J | F | M | A | M | J |
| J | A | S | O | N | D |

Norway Spruce
Picea abies subsp. *abies*

ID FACT FILE

CROWN:
Conical

BARK:
Reddish-brown,
smooth at first,
becoming finely
cracked

LEAVES:
Narrow, 4-sided
needles,
10–25 mm, stiff,
sharp-pointed,
dark green;
spreading to
reveal lower side
of shoot; persis-
tent peg-like
bases

CONES:
10–18 cm, erect
and dark red at
first but
red-brown and
pendulous when
mature. Cone
scales have
squarish or
notched tips

Norway Spruce is a major forest tree throughout much of N Europe and in mountains as far south as the Alps and the Balkan Peninsula. It is also planted for timber and to provide shelter. In Britain and elsewhere it is perhaps most familiar as the Christmas tree; many young trees are grown each year specifically for this purpose.

PINE FAMILY, PINACEAE

Evergreen
Up to 55 m

J	F	M	A	M	J
J	A	S	O	N	D

Douglas Fir
Pseudotsuga menziesii

ID FACT FILE

CROWN:
Conical. Branches in irregular whorls

BARK:
Grey to purple-brown, corky and ridged

LEAVES:
Narrow needles 20–35 mm, pointed but soft, dark green with 2 white bands beneath, aromatic when crushed; spreading to sides of shoot. Leaf-scars raised, elliptical

CONES:
5–10 cm, brown, pendulous. A long, 3-toothed bract projects beneath each scale

LOOKALIKES:
Spruces

In its native western N America some specimens of Douglas Fir are among the tallest trees in the world, exceeding 100 m in height. Trees grown in Europe flourish in damp regions but rarely reach anywhere near this size. Widely planted, Douglas Fir is easily recognised by the cones which have long, three-toothed bracts beneath the cone-scales. It sometimes exceeds its normal maximum height of 55 m.

PINE FAMILY, PINACEAE

Evergreen
Up to 35 m

J	F	M	A	M	J
J	A	S	O	N	D

Scots Pine
Pinus sylvestris

ID FACT FILE

CROWN:
Small,
flat-topped,
usually irregular

BARK:
Reddish–brown,
paler and papery
towards top of
trunk

LEAVES:
Paired needles
25–80 mm,
twisted, finely
toothed, grey or
blue-green with a
grey sheath at
base of each pair

CONES:
2–8 cm, in
clusters of 1–3,
ripening to dull
grey in second
year. Scales
have a flat or
pyramidal apex
and a short
spine

Scots Pine is easily recognised by its distinctive long, bare, reddish-brown trunk and small, lop-sided crown. It is widespread in Europe and is the only species of pine native to Britain. Extensive natural pine forests are now restricted to Scotland but small woods and individual trees occur in many areas and Scots Pine is widely planted for timber.

PINE FAMILY, PINACEAE

Deciduous
Up to 35 m

| J | F | M | A | M | J |
| J | A | S | O | N | D |

European Larch
Larix decidua

Larches are deciduous conifers with shoots of two kinds; long shoots with scattered needles and short spur-like shoots with rosettes of 30–40 needles. The needles turn red and then yellow in autumn before falling. European Larch is a short-lived, fast-growing tree native to mountains of central and Eastern Europe but widely planted for timber and naturalised in many places.

ID FACT FILE

CROWN:
Conical, becoming broader. Shoots pendulous

BARK:
Grey to brownish, cracked

LEAVES:
Flattened needles 12–30 mm, soft, pale green, scattered on long shoots, in dense rosettes on spur-like short shoots

CONES:
2–3 cm, ovoid. Scales softly hairy and close-pressed

CYPRESS FAMILY, CYPERACEAE

Evergreen
Up to 6 m

J	F	M	A	M	J
J	A	S	O	N	D

Juniper
Juniperus communis

ID FACT FILE

CROWN:
Spreading, bushy

BARK:
Reddish,
shredding

LEAVES:
Prickly, flattened
needles in
whorls of 3;
8–30 mm, stiff,
bluish-green,
upper side with
broad white band

CONES:
Fleshy and
berry-like,
6–9 mm, globu-
lar, ripening dull
blue-black in
second or third
year

Junipers form small trees or shrubs on
lime-rich soils in most parts of Europe. They
are typical of downland in the north but are
confined to mountainous regions in the south.
Male and female cones are borne on separate
trees. Initially the female cones resemble those
of cypresses but as they ripen the scales
become fleshy and coalesce, forming a
berry-like structure instead of a woody cone.

WILLOW FAMILY, SALICACEAE

Deciduous
Up to 7 m

J	F	M	A	M	J
J	A	S	O	N	D

Bay Willow

Salix pentandra

ID FACT FILE

CROWN:
Bushy

TWIGS:
Shiny reddish-brown

LEAVES:
Alternate, 5–12 cm, less than 3 times as long as wide, elliptical, shiny above

FLOWERS:
Catkins appear with leaves

Bay Willow is widespread in N and central Europe but is native only to some parts. Male trees have more showy catkins than female trees and are more often planted. In areas where the species is merely introduced, most or all of the trees may be male. The leaves supposedly have a similar scent to those of the Sweet Bay though in fact they smell faintly of balsam.

WILLOW FAMILY, SALICACEAE

Deciduous
Up to 10 m

J	F	M	A	M	J
J	A	S	O	N	D

Grey Sallow
Salix cinerea

ID FACT FILE

CROWN:
Broad, with branches spreading

TWIGS:
Thickly grey-hairy when young

LEAVES:
Alternate, 2–16 cm, less than 3 times as long as wide, grey-hairy beneath, toothed margins rolled under. Ear-shaped stipules at base

FLOWERS:
Catkins erect, appearing before leaves

Also called Grey Willow, this tree is named for the ash-coloured hairs which densely cover the young twigs and the undersides of the leaves. The silky-hairy catkins are also grey and, like Goat Willow, are often referred to as 'pussy willows'. A subspecies of Grey Sallow with red-brown twigs and leaves with rusty-coloured hairs beneath is separated as the Rusty Sallow.

WILLOW FAMILY, SALICACEAE

Deciduous
Up to 6 m

J	F	M	A	M	J
J	A	S	O	N	D

Osier
Salix viminalis

ID FACT FILE

CROWN:
Narrow

TWIGS:
Long, grey-hairy
becoming shiny
olive or brown

LEAVES:
Alternate,
10–15 cm, tape-
ring, margins
often wavy or
rolled under,
silvery-hairy
beneath

FLOWERS:
Catkins erect,
crowded at twig
tips, appearing
before leaves

Osier is common almost everywhere in wet,
lowland habitats but is probably introduced in
many areas. It is a favourite species for
basket-making and is often planted to form
osier beds. Here the trees are pollarded to
leave a short trunk with a rounded head of
long, pliant twigs which are regularly cropped.
These twigs are the 'withies' used for baskets,
lobster-pots and other cane-work.

WILLOW FAMILY, SALICACEAE

Deciduous
Up to 20 m

J	F	M	A	M	J
J	A	S	O	N	D

White Poplar
Populus alba

White Poplar readily produces suckers which spread underground for a considerable distance before emerging and which often form thickets around the parent tree. The leaves are two-coloured, the hairy, white lower surface contrasting strikingly with the hairless, dark green upper surface. White Poplar is native to Europe though it is introduced in many areas. It prefers soft, wet ground.

ID FACT FILE

CROWN:
Spreading

BARK:
Grey with black bars

LEAVES:
Alternate, 3–9 cm, with irregular lobes, densely white-hairy beneath. Stalk cylindrical

FLOWERS:
Catkins appear before leaves. Males purplish, females greenish

WILLOW FAMILY, SALICACEAE

Deciduous
Up to 20 m

J	F	M	A	M	J
J	A	S	O	N	D

Aspen
Populus tremula

ID FACT FILE

CROWN:
Broad, some-
times conical

BARK:
Grey, smooth

LEAVES:
Alternate,
1.5–8 cm, bluntly
toothed, very
pale beneath.
Stalk flattened
on sides

FLOWERS:
Catkins appear
before leaves;
males purplish,
females pinkish

This is the most widespread of the European
poplars and is found almost throughout the
region. Aspens are renowned for the
near-continuous motion of their 'trembling'
leaves. This is due to the lateral flattening of
the leaf-stalks which allows the leaves to flutter
in the faintest air currents. The movement is
accentuated by the flashing of the pale lower
surfaces of the leaves.

WILLOW FAMILY, SALICACEAE

Deciduous
Up to 35 m

J	F	M	A	M	J
J	A	S	O	N	D

Black Poplar
Populus nigra

ID FACT FILE

CROWN:
Broad, rounded.
Trunk has rough
swellings

BARK:
Grey, fissured

LEAVES:
Alternate,
5–10 cm, finely
toothed. Stalk
flattened on
sides

FLOWERS:
Catkins appear
before leaves;
males crimson,
females greenish

Black Poplar forms a robust tree, the trunk often with large swellings and numerous twiggy outgrowths. Unlike those of other common European species the finely toothed leaves are smooth and only slightly paler on the lower surface than on the upper. The leaves have laterally flattened stalks and are able to flutter like those of Aspen but the movement is less pronounced.

BIRCH FAMILY, BETULACEAE

Deciduous
Up to 20 m

J	F	M	A	M	J
J	A	S	O	N	D

Common Alder
Alnus glutinosa

ID FACT FILE

CROWN:
Broadly conical.
Young twigs
sticky

BARK:
Grey

LEAVES:
Alternate,
4–10 cm.
Margins doubly
toothed

FLOWERS:
Catkins appear
before leaves.
Males 2–6 cm,
pendulous;
females 1.5 cm,
ovoid, stalked, in
clusters

FRUITS:
Woody, cone-like,
1–3 cm

Common Alders are typically found on marshy soils along streams and rivers and may be the dominant trees in wet places. The roots harbour nitrogen-fixing bacteria which enable the trees to thrive in such soils. The woody fruits superficially resemble the cones of pines and other conifers but have a completely different anatomical structure.

BIRCH FAMILY, BETULACEAE

Deciduous
Up to 25 m

| J | F | M | A | M | J |
| J | A | S | O | N | D |

Downy Birch
Betula pubescens

ID FACT FILE

CROWN:
Branches erect
to spreading

BARK:
Brown or grey.
Smooth base

LEAVES:
Alternate, up to
5.5 cm with
rounded or trian-
gular base,
coarsely toothed

FLOWERS:
Male catkins
3–6 cm, pendu-
lous at tips of
twigs; females
1–4 cm in leaf
axils

FRUITS:
Tiny, winged nut-
lets released
when female
catkins break up

A short-lived, cold-tolerant species widespread
in Europe, especially in the north and in
mountains where it forms extensive forests.
Normally a small tree, in harsh or exposed
areas it grows as a shrub. The male catkins are
formed at the tips of twigs at the beginning of
winter, ready to mature and shed pollen the
following spring.

BIRCH FAMILY, BETULACEAE

Deciduous
Up to 30 m

J	F	M	A	M	J
J	A	S	O	N	D

Silver Birch
Betula pendula

ID FACT FILE

CROWN:
Slender. Branches pendulous at tips

BARK:
Silvery

LEAVES:
Alternate, up to 5 cm with base heartshaped or cut straight across, doubly toothed

FLOWERS:
Male catkins at tips of twigs; females in leaf axils

FRUITS:
Tiny, winged nutlets

An elegant, slender tree with distinctive silvery-white bark. The bark is mostly smooth but breaks up into darker, rectangular plates at the base of the trunk. A fast-growing, short-lived species. Groves of saplings may spring up as early colonisers where other trees have fallen or been cut down before becoming shaded out by taller, slower-growing species.

HAZEL FAMILY, CORYLACEAE

Deciduous
Up to 12 m

| J | F | M | A | M | J |
| J | A | S | O | N | D |

Hazel
Corylus avellana

ID FACT FILE

CROWN:
Shrubby,
spreading

BARK:
Smooth, peeling
in thin, horizontal
strips

LEAVES:
Alternate, up to
10 cm, very
broadly oval,
doubly toothed,
rough and bristly
to the touch

FLOWERS:
Appearing before
leaves; males in
catkins, females
very small, in
clusters

FRUITS:
Nut enveloped in
green, leafy cup

A very common hedgerow and woodland
species throughout Europe. It often grows as a
shrub and is frequently coppiced. Hazels are
among the first trees to flower in spring, the
bright yellow male catkins hanging from the
bare branches as early as January. The
hard-shelled nuts are each surrounded by a
ragged-tipped, leafy cup or involucre and
contain an edible kernel.

BEECH FAMILY, FAGACEAE

Deciduous
Up to 30 m

| J | F | M | A | M | J |
| J | A | S | O | N | D |

Sweet Chestnut
Castanea sativa

ID FACT FILE

CROWN:
Tall, uneven

BARK:
Greyish, spirally
grooved

LEAVES:
Alternate,
10–25 cm,
oblong, pointed,
sharp-toothed

FLOWERS:
Erect catkins
with yellow male
flowers in upper
part, green
females below

FRUITS:
Shiny, red-brown
nuts enclosed in
spiny husk

Although native to S Europe, Sweet Chestnut
has been widely planted elsewhere for its
edible nuts. Between one and three nuts are
enclosed in a rather softly but densely spiny
husk which splits while still on the tree to
release them. The Sweet Chestnut should not
be confused with the unrelated horse-
chestnuts which have fewer, tougher spines
and nuts which are inedible.

BEECH FAMILY, FAGACEAE

Deciduous
Up to 40 m

| J | F | M | A | M | J |
| J | A | S | O | N | D |

Beech
Fagus sylvatica

ID FACT FILE

CROWN:
Broadly domed

BARK:
Smooth, grey

LEAVES:
Alternate, oval to
elliptical, edges
wavy

FLOWERS:
Males in droop-
ing, long-
stalked heads;
females paired

FRUITS:
Spiny, 4-lobed
husk containing
2 small, triangu-
lar nuts

Beech woods are characteristic of limestone
regions. The leaves are very slow to rot and
improve the soil, and in pure beech woods a
deep layer of dead leaves builds up on the
ground. This prevents other plants from
growing well, so beech woods have few
woodland flowers. Beech nuts, or mast, are
much-loved by pigs which were often turned
into the woods to forage in autumn.

BEECH FAMILY, FAGACEAE

Deciduous
Up to 45 m

| J | F | M | A | M | J |
| J | A | S | O | N | D |

Pedunculate Oak
Quercus robur

ID FACT FILE

CROWN:
Very large,
spreading

BARK:
Dark grey, deeply
furrowed

LEAVES:
Alternate,
10–12 cm, lobes
irregular, in 5–7
pairs

FRUITS:
Acorns in clus-
ters on a long
stalk. Acorn cup
11–18 mm
across. All but
tips of cup-
scales fused
together

Sometimes called English Oak, this species is typically the dominant tree of deciduous woodlands, especially on heavy clay soils in lowland regions. It is a massive tree with a broad, spreading crown and can live to a great age, up to 800 years. The first growth of leaves is often badly attacked by insects and a second growth, appearing reddish when young, may be produced in summer.

BEECH FAMILY, FAGACEAE

Deciduous
Up to 35 m

| J | F | M | A | M | J |
| J | A | S | O | N | D |

Red Oak
Quercus rubra

ID FACT FILE

Crown:
Broadly domed.

Twigs:
Dark red

Bark:
Silvery, smooth

Leaves:
Alternate,
12–22 cm, lobes
reaching half-way
to middle and
with slender
teeth

Fruits:
Acorn cup very
shallow,
15–25 mm
across.
Cup-scales thin
with fine hairs

This eastern N American species is now widely grown in Europe for its splendid autumn colours. The leaves, which have lobes tipped with slender teeth, turn a dark, vibrant red in autumn and contrast with the silvery colour of the bark. The best colours are found in young trees which are often grown planted along roads; old trees produce more yellows and browns.

ELM FAMILY, ULMACEAE

Deciduous
Up to 40 m

| J | F | M | A | M | J |
| J | A | S | O | N | D |

Wych Elm
Ulmus glabra

ID FACT FILE

CROWN:
Very broad,
spreading

BARK:
Greyish, smooth,
becoming ridged

LEAVES:
Alternate,
10–18 cm, stiffly
hairy above,
softer beneath,
unequal base
has one side
curved over stalk

FLOWERS:
Purplish clusters
appearing before
leaves

FRUITS:
Papery discs
15–20 mm long
with a centrally
placed seed

Elms are easily recognised by their papery, winged seeds and by their leaves in which the blade extends further down the midrib on one side than on the other. Wych Elm is native to much of Europe and is probably the commonest species remaining in Britain following the outbreak of Dutch Elm disease to which it is slightly resistant. However it does not produce suckers so affected trees cannot regenerate.

ELM FAMILY, ULMACEAE

Deciduous
Up to 30 m

| J | F | M | A | M | J |
| J | A | S | O | N | D |

Small-leaved Elm
Ulmus minor

ID FACT FILE

CROWN:
Narrow,
branches angled
upwards

TWIGS:
Often with
well-developed,
corky wings

LEAVES:
Alternate,
6–8 cm, smooth
on both sides,
long side of
unequal base
turned abruptly
to join stalk

FLOWERS:
Reddish clusters
appearing before
leaves

FRUITS:
Papery disc
7–18 mm long
with seed placed
near top edge

Small-leaved Elm is native throughout much of
Europe and is a very variable species. Local
populations of trees are sufficiently distinct
from each other to be given different names
such as Cornish Elm, Coritanian Elm and
Jersey Elm. They are sometimes regarded as
forming separate species but are very difficult
to tell apart.

ROSE FAMILY, ROSACEAE

Deciduous
Up to 20 m

J	F	M	A	M	J
J	A	S	O	N	D

Common Pear
Pyrus communis

Common Pear was probably first introduced to
Europe from W Asia and is now naturalised
and widespread in woods and hedgerows.
Solitary trees, however, are often the only surviving remnants of the gardens of old, abandoned houses. All of the orchard trees grown
for their edible fruits, of which there are many
different types, belong to the variety *culta*.

ID FACT FILE

CROWN:
Narrow

TWIGS:
Spiny in older
trees

LEAVES:
Alternate,
5–8 cm, toothed,
densely hairy at
first

FLOWERS:
In clusters
appearing with
the leaves; 5
petals

FRUITS:
4–12 cm,
pear-shaped to
globose,
sweet-tasting or
sour

ROSE FAMILY, ROSACEAE

Deciduous
Up to 10 m

J	F	M	A	M	J
J	A	S	O	N	D

Crab Apple
Malus sylvestris

ID FACT FILE

CROWN:
Dense

TWIGS:
Often spiny

LEAVES:
Alternate,
3–11 cm, hairless on both sides

FLOWERS:
3–4 cm across, in clusters. Persistent sepals hairy on inner surface; 5 white or pink petals

FRUITS:
2.5–3 cm, hard, sour

Crab Apples are the ancestors of the cultivated apples. They were themselves once domesticated as fruit trees and introduced to many areas before falling out of use. The hard, sour fruits are still sometimes used for jellies and jams. Truly wild trees are spiny and have almost white flowers. The descendants of domesticated trees are unarmed and have pinkish flowers.

ROSE FAMILY, ROSACEAE

Deciduous
Up to 18 m

J	F	M	A	M	J
J	A	S	O	N	D

Hawthorn
Crataegus monogyna

ID FACT FILE

BARK:
Grey to brown,
cracking into
plates

TWIGS:
Numerous thorns
up to 15 mm

LEAVES:
Alternate, shiny,
1.5–4.5 cm,
ovate but deeply
divided into usu-
ally 5–7 lobes

FLOWERS:
8–15 mm
across, white or
pink with 1 style

FRUITS:
7–14 mm, dark
or bright red; 1
seed

Hawthorns are perhaps best known as
hedgerow shrubs and have been used for this
purpose since hedges were first planted in
Europe. They provide a quick-growing barrier
to animals and their thorny twigs protect them
from browsing. Hawthorn hedges come into
leaf earlier and are denser than other hedges
and support a greater variety of wildlife than
any other. Left uncut, Hawthorns form sturdy,
densely crowned trees.

ROSE FAMILY, ROSACEAE

Deciduous
Up to 10 m

| J | F | M | A | M | J |
| J | A | S | O | N | D |

Juneberry
Amelanchier lamarkii

ID FACT FILE

CROWN:
Slender, open

TWIGS:
Shaggily white-hairy when young

LEAVES:
Alternate, up to 8 cm, oblong to elliptical. Margins slightly upturned

FLOWERS:
In clusters, stalks hairy. Petals white, narrow and erect

FRUITS:
1 cm, purplish-black with pale bloom, crowned with withered sepals

Juneberries are a group of mainly American shrubs and small trees. They produce drifts of white blossom in spring followed by edible fruits which appear, appropriately, from June to about August. They are much sought after by animals, birds and humans. This species is thought to be a natural hybrid which arose in the wild and was later introduced to Europe where it is widely planted for ornament and is naturalised in places.

ROSE FAMILY, ROSACEAE

Deciduous
Up to 20 m

J	F	M	A	M	J
J	A	S	O	N	D

Rowan
Sorbus aucuparia

ID FACT FILE

BUDS:
Pointed, purple

LEAVES:
Alternate,
pinnate, leaflets
in 5–10 pairs,
3–6 cm, towards
the tip, grey-hairy
beneath

FLOWERS:
White, in
flat-topped
clusters

FRUITS:
6–9 mm,
globose, scarlet

Sometimes called Mountain Ash, Rowan is
common in upland areas, though it will grow at
any altitude. Formerly used as a charm against
many forms of witchcraft, its major modern
use is as an attractive ornamental whose
narrow shape makes it ideal as a street tree. Its
fruits are avidly eaten by birds and will attract
them even into town centres in winter.

ROSE FAMILY, ROSACEAE

Deciduous
Up to 10 m

| J | F | M | A | M | J |
| J | A | S | O | N | D |

Wild Plum
Prunus domestica

ID FACT FILE

BRANCHES:
Straight, some-
times thorny

LEAVES:
Alternate,
3–8 cm, oval,
toothed, dull
green, smooth
above, downy
beneath

FLOWERS:
In clusters of
2–3, appearing
with leaves; 5
white petals

FRUITS:
2–7.5 cm, colour
ranging from
greenish to red,
purple or
blue-black, some-
times with a
waxy bloom

A complex hybrid between Blackthorn and
Cherry Plum, Wild Plum includes several
different fruit trees. The cultivated Plum has
sparsely hairy and spineless twigs and large
fruits. The cultivated Damson and Greengage
and the wild Bullace all have densely hairy,
spiny twigs and smaller fruits, those of Damson
purple, those of Greengage yellow and those of
Bullace red.

ROSE FAMILY, ROSACEAE

Deciduous
Up to 30 m

J	F	M	A	M	J
J	A	S	O	N	D

Wild Cherry
Prunus avium

Wild Cherry grows rapidly and attains a
greater size than most other species of cherry.
It is found wild in mixed and deciduous woods
throughout most of Europe, especially in hilly
regions, and is also grown for its fruit and
timber. The fruits are usually dark red but
range from yellowish to bright red or black and
may be sweet or sour when ripe.

ID FACT FILE

CROWN:
High, domed

BARK:
Red-brown, pee-
ling in horizontal
bands

LEAVES:
Alternate,
8–15 cm, oval to
oblong, pointed,
with sharp,
forward-pointing
teeth, veins with
hairy tufts in
angles

FLOWERS:
In clusters of
2–6, appearing
just before
leaves; 5 petals,
1–1.5 cm, white

FRUITS:
Up to 9–12 mm,
usually dark red

ROSE FAMILY, ROSACEAE

Deciduous
Up to 30 m

| J | F | M | A | M | J |
| J | A | S | O | N | D |

Rum Cherry
Prunus serotina

ID FACT FILE

CROWN:
Spreading

BARK:
Grey, smooth but peeling, aromatic

LEAVES:
Alternate, 5–14 cm, tip tapering, fine teeth forward-pointing, shiny above, slightly downy beneath

FLOWERS:
In spikes 7–15 cm long; 5 petals, 3–5 mm, white, margins minutely toothed

FRUITS:
6–8 mm, glossy black, sepals persisting at tip

Rum Cherry is a rather stout tree native to eastern N America. In Europe it is grown mainly for timber and occasionally as an ornamental. The bark is less colourful than in other cherries and has a bitter, aromatic scent. The bitter fruits are also unusual in being crowned with the withered remains of the persistent sepals, like an apple. Most cherries are completely smooth.

PEA FAMILY, LEGUMINOSAE

Deciduous
Up to 25 m

| J | F | M | A | M | J |
| J | A | S | O | N | D |

False Acacia
Robinia pseudoacacia

ID FACT FILE

CROWN:
Open

BARK:
With spiral ridges

LEAVES:
Alternate,
15–20 cm, pin-
nate, yellowish-
green leaflets in
3–10 pairs. Stalk
has 2 spines at
base

FLOWERS:
Pea-like, white,
fragrant, in han-
ging clusters

FRUITS:
Pods 5–10 cm

Native to N America but a popular garden and
street tree in Europe. It may produce several
trunks if the suckers which readily appear
around the base are allowed to grow. Despite
being large and showy, the flower clusters are
often produced quite high in the crown and
are easy to miss. However, the pods remain on
the branches long after the leaves have fallen.

MAPLE FAMILY, ACERACEAE

Deciduous
Up to 35 m

| J | F | M | A | M | J |
| J | A | S | O | N | D |

Sycamore
Acer pseudoplatanus

ID FACT FILE

CROWN:
Wide-spreading

LEAVES:
Opposite,
10–15 cm, the 5
spreading lobes
coarsely toothed

FLOWERS:
Appearing with
the leaves,
yellowish-green.
Males and
females in sepa-
rate, hanging
clusters

FRUITS:
Joined at the
base in pairs
3.5–5 cm
long, the wings
forming a right-
angle

Sycamores are fast-growing, invasive trees
which rapidly colonise new areas. The winged
seeds can be carried over considerable
distances by the wind and they germinate
easily. The leaves of rapidly growing young
trees are deeply lobed, those of old,
slower-growing trees are more shallowly divided. The largest of the European maples, and
native or naturalised in most areas.

Deciduous
Up to 25 m

J	F	M	A	M	J
J	A	S	O	N	D

Field Maple
Acer campestre

ID FACT FILE

CROWN:
Rounded

TWIGS:
Often with corky
wings

LEAVES:
Opposite,
4–12 cm,
3 lobes them-
selves further
lobed or with
rounded teeth
towards tips

FLOWERS:
Yellowish-green,
in erect clusters
appearing with
leaves; 5 petals

FRUITS:
Paired, 2–4 cm
long, wings
horizontal

A common tree in N Europe. Many of the largest specimens have been felled for their timber, known as bird's-eye maple, and mostly smaller specimens remain. Trees that are pruned regularly, such as those growing in hedgerows, may develop broad, corky wings along the twigs. Despite the leathery texture of the leaves, Field Maple is deciduous, the leaves turning butter-yellow in autumn before falling.

HORSE-CHESTNUT FAMILY, HIPPOCASTANACEAE

Deciduous
Up to 35 m

| J | F | M | A | M | J |
| J | A | S | O | N | D |

Horse-chestnut
Aesculus hippocastanum

ID FACT FILE

CROWN:
Wide-spreading

TWIGS:
Red-brown, with large, sticky winter buds

LEAVES:
Opposite, palmate; 5–7 toothed leaflets each 10–25 cm long

FLOWERS:
2 cm across, in erect, conical spikes; 4–5 petals, frilly, stamens protruding

FRUITS:
Up to 6 cm, spiny husk enclosing shiny brown seeds

This large, spreading tree is native to the Balkan mountains but has been introduced to many other parts of Europe. It is commonly planted in broad avenues and parks. The spiny fruits contain 1–3 shiny brown seeds – the familiar conkers – which germinate readily, so the tree is often naturalised. Horse-chestnut is easily recognised in winter by its very sticky buds which may be up to 3.5 cm long.

HOLLY FAMILY, AQUIFOLIACEAE

Evergreen
Up to 15 m

| J | F | M | A | M | J |
| J | A | S | O | N | D |

Holly
Ilex aquifolium

ID FACT FILE

CROWN:
Conical

BARK:
Silver-grey,
smooth

LEAVES:
Alternate,
5–12 cm, waxy,
stiff and
leathery. Wavy
margins with
spines

FLOWERS:
Up to 6 mm, in
small clusters;
males and
females on sepa-
rate trees;
4 petals white

FRUITS:
Berries 7–12 mm
diameter, scarlet

Holly is very tolerant of shade and often forms
a shrubby understorey in woods where other
small trees cannot survive. An old tradition that
it was unlucky to cut hollies means that many
are still seen growing as single trees in
otherwise well-trimmed hedgerows. Berries
are only produced by female trees and are a
favourite winter food for birds, especially
thrushes.

SPINDLE FAMILY, CELASTRACEAE

Deciduous
Up to 6 m

| J | F | M | A | M | J |
| J | A | S | O | N | D |

Spindle Tree
Euonymus europaeus

ID FACT FILE

TWIGS:
4-angled when
young

LEAVES:
Opposite,
3–10 cm, oval
with tapering tip.
Margins toothed

FLOWERS:
In inconspicuous
clusters; 4
narrow, greenish-
yellow petals

FRUITS:
Pink, 4-lobed
capsule
1–1.5 cm

Spindle Trees are common on lime-rich soils
but are inconspicuous for much of the year as
the leaves have few distinguishing features and
the flowers are small and greenish. They are
much more noticeable in autumn when the
leaves turn dark red and the fruits ripen. The
peculiarly shaped capsules have four distinct
lobes and are matt pink. Each lobe splits to
reveal a single orange seed.

BUCKTHORN FAMILY, RHAMNACEAE

Deciduous
Up to 10 m

J	F	M	A	M	J
J	A	S	O	N	D

Buckthorn
Rhamnus catharticus

Buckthorn produces two kind of shoots: long shoots which bear widely separated pairs of leaves and extend the tree's growth, and short side-shoots which bear crowded leaves and the inconspicuous flower-clusters. Dense clusters of attractive-looking black berries are produced by female trees. They are a violent laxative, hence the tree's alternative name, Purging Buckthorn.

ID FACT FILE

BRANCHES:
At right-angles

LEAVES:
Opposite, crowded on short side-twigs which end in a spine, 3–7 cm, oval with 2–4 pairs of curved veins

FLOWERS:
In small clusters, fragrant, males and females on separate trees; 4 narrow, greenish petals

FRUITS:
Berries 6–8 mm in diameter, black

BUCKTHORN FAMILY, RHAMNACEAE

Deciduous
Up to 5 m

J	F	M	A	M	J
J	A	S	O	N	D

Alder Buckthorn
Frangula alnus

A small tree, often no more than a shrub,
found on acid soils. The leaves characteristically
hang downwards on the twigs as they change
colour in autumn. The berries are green when
immature and turn yellow, then red before
ripening to black. Despite its name, Alder
Buckthorn is not related to Alder but does
grow in similar, marshy conditions. It is
common in much of Europe.

ID FACT FILE

BRANCHES:
Opposite, angled
upwards

LEAVES:
Mostly opposite,
2–7 cm, widest
above middle,
with 7–9 pairs of
curved veins,
margins entire,
wavy

FLOWERS:
Small, in clus-
ters; 5 greenish-
white petals

FRUITS:
Berries 6–10 mm
in diameter,
purple-black

OLEASTER FAMILY, ELAEAGNACEAE

Deciduous
Up to 11 m

| J | F | M | A | M | J |
| J | A | S | O | N | D |

Sea Buckthorn
Hippophae rhamnoides

ID FACT FILE

BARK:
Blackish

TWIGS:
Thorny, silvery
when young

LEAVES:
Alternate, 1.6 cm
long but only
3–10 mm wide.
Silvery on both
sides or dull
grey-green above

FLOWERS:
Greenish,
appearing before
leaves; males
and females on
separate trees;
tubular with 2
sepals, petals
absent

FRUITS:
Berries 6–8 mm,
oval, orange

Sea-buckthorn is confined to seaside places, on cliffs, dunes and sand-bars. In these exposed sites it seldom reaches its full height, instead forming sprawling shrubs stunted by the wind. All parts of the tree are covered with minute, silvery scales which are easily rubbed off. These are too small to see with the naked eye but give a distinctive silvery cast to the whole tree.

LIME FAMILY, TILIACEAE

Deciduous
Up to 32 m

J	F	M	A	M	J
J	A	S	O	N	D

Small-leaved Lime
Tilia cordata

ID FACT FILE

CROWN:
Dense. Branches
arching down-
wards

LEAVES:
Alternate,
3–9 cm,
heart-shaped,
finely toothed,
pale green
beneath with
reddish tufts of
hairs in angles of
veins

FLOWERS:
White, fragrant,
hanging in a
cluster beneath
a pale greenish
wing

FRUITS:
Ribbed nuts
6 mm across

Native to limestone areas and chalky soils in
most of Europe except the far north and south.
Small-leaved Lime was probably the last tree
species to reach Britain after the Ice-Age but
before the formation of the English Channel.
It was a characteristic tree of the original
forests in lowland Britain but did not reach
Ireland and cannot grow from seed in the
cooler climate of Scotland.

LIME FAMILY, TILIACEAE

Deciduous
Up to 40 m

| J | F | M | A | M | J |
| J | A | S | O | N | D |

Large-leaved Lime
Tilia platyphyllos

This is the earliest-flowering of the Limes, producing blossom several weeks before other species. The flowers produce large quantities of nectar and bees, in particular, swarm around the trees, attracted to this rich food source. Sugars in the nectar can be poisonous in large quantities and the trees are often surrounded by dead or dying bees, especially bumble-bees.

ID FACT FILE

CROWN:
Narrow, branches angled upwards

LEAVES:
Alternate, 6–9 cm, heart-shaped, sharply toothed, hairy on both surfaces

FLOWERS:
Yellowish, fragrant, hanging in a cluster beneath a greenish-white wing

FRUITS:
Nuts globular, 8–12 mm across with 3–5 ribs

DOGWOOD FAMILY, CORNACEAE

Deciduous
Up to 8 m

J	F	M	A	M	J
J	A	S	O	N	D

Cornelian Cherry
Cornus mas

Copious blossom appears before the leaves and turns this tree into a mass of yellow in late winter or early spring. When the leaves do appear they are also a distinctive yellowish-green with very prominent veins. The fruits resemble rather oblong, short-stalked cherries and are edible but somewhat tart, even when ripe. They are usually made into preserves.

ID FACT FILE

CROWN:
Spreading

LEAVES:
Opposite,
4–10 cm, oval to elliptical with prominent veins, yellowish-green

FLOWERS:
Appearing before leaves in clusters 2 cm across; 4 bright yellow petals

FRUITS:
Berries cherry-like but rather oblong, 12–20 mm, bright red

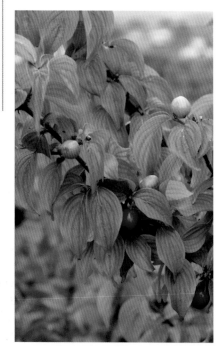

OLIVE FAMILY, OLEACEAE

Deciduous
Up to 40 m

J	F	M	A	M	J
J	A	S	O	N	D

Common Ash
Fraxinus excelsior

A common and widespread tree of woodlands and hedgerows, preferring chalky or lime-rich soils. It is one of the later trees to come into leaf in the spring. Tiny flowers are wind-pollinated and lack sepals and petals. Males and females may be borne on separate trees or on separate branches of the same tree. Easy to identify in winter by the prominent black buds in opposite pairs.

ID FACT FILE

CROWN:
Domed, open

BARK:
Grey, smooth, becoming ridged with age

WINTER BUDS:
Black

LEAVES:
Opposite, pinnate, 7–13 leaflets each 5–12 cm, toothed. Midrib with white hairs beneath

FLOWERS:
Tiny, purplish, appearing before leaves in sepa-rate male and female clusters

FRUITS:
2.5–5 cm, winged

HONEYSUCKLE FAMILY, CAPRIFOLIACEAE

Deciduous
Up to 10 m

| J | F | M | A | M | J |
| J | A | S | O | N | D |

Elder
Sambucus nigra

ID FACT FILE

CROWN:
Branches curved
outwards

BARK:
Pale brown, corky
and deeply
grooved

LEAVES:
Opposite,
pinnate, 5–7
leaflets each
4.5–12 cm,
sharply toothed

FLOWERS:
Fragrant, numer-
ous in branched,
flat-topped clus-
ters 10–24 cm
across; 5 white
petals

FRUITS:
Berries 6–8 mm,
black, whole
cluster drooping
when ripe

Elders are common on disturbed soil which is
rich in nitrogen and grow quickly to form large
shrubs or small trees. The leaves are rank
smelling but the flowers are cloyingly sweet.
The edible berries which follow are ripe when
the heads droop. Both flowers and berries are
rich in vitamin C and are collected for
domestic and commercial use. All other parts
of the plant are mildly poisonous.

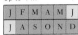

Deciduous
Up to 4 m

J	F	M	A	M	J
J	A	S	O	N	D

Guelder Rose
Viburnum opulus

ID FACT FILE

TWIGS:
Angled, greyish, hairy

LEAVES:
Opposite, 3–8 cm, with 3–5 spreading, toothed lobes

FLOWERS:
White, in circular heads 4.5–10.5 cm across with large sterile flowers around the rim and small fertile flowers in the centre; 5 petals

FRUITS:
8 mm, scarlet, translucent

LOOKALIKES:
Maples

Found throughout most of Europe, Guelder Rose prefers damp habitats and is one of the few tree species to thrive in fenlands. The flowerheads have a ring of large, showy outer flowers surrounding much smaller and more numerous inner ones. The outer flowers are sterile but serve to attract pollinating insects to the fertile inner flowers. The bark, leaves and berries are all poisonous.

AMANITACEAE

J	F	M	A	M	J
J	A	S	O	N	D

Fly Agaric
Amanita muscaria

ID FACT FILE

CAP: 10–15 cm, convex, scarlet to orange-red with fluffy white scales

STEM: To 20 × 2 cm, white, with ring and bulbous, scaly base

GILLS: White, free. Spores white

HABITAT: Under birch or pine

SEASON: Late summer to late autumn

LOOKALIKES: Unmistakable when fresh

A common species in birch and pine woods, easily recognised in fresh condition. However, the bright red colours may fade to orange-yellow in old, weathered specimens and cap scales may be lost. The veil is fragile, leaving scale-like remains at the stem base. A poisonous species, having hallucinogenic effects; known as the Sacred Mushroom, and of great significance in the folklore of many parts of Europe. It contains ibotenic acid, a weak insecticide, and since medieval times it has been used as a fly killer.

AMANITACEAE

J	F	M	A	M	J
J	A	S	O	N	D

Death Cap
Amanita phalloides

ID FACT FILE

CAP: 5–12 cm, convex, olive-green, radially streaky, smooth or with white veil remnants

STEM: White, with large ring when fresh, bulbous base and large volva

GILLS: Free, whitish. Spores white

HABITAT: Beech and oak woods

SEASON: Late summer–late autumn

LOOKALIKES: False Death Cap is yellowish and lacks large volva. Edible mushrooms have dark gills at maturity

This species is deadly poisonous and is responsible for most deaths from fungal poisoning in Europe. It contains cyclopeptides, cell-damaging compounds, the most dangerous of which are amatoxins. These are not broken down in cooking and cause severe liver damage. Primary symptoms, caused by phallotoxins, appear 8–15 hours after ingestion; those caused by the ama-toxins occur later. Appears in oak woods, and can be recognised by the greenish, streaky cap, large ring and volva. White forms also occur.

AMANITACEAE

J	F	M	A	M	J
J	A	S	O	N	D

ID FACT FILE

Cap: 6–12 cm, convex, reddish-brown with grey-brown, scale-like remains of veil

Stem: Whitish or reddish-brown, with ring and swollen, scaly base

Gills: Free, white, often with red-brown spots. Spores white

Habitat: Deciduous and coniferous woods

Season: Summer–late autumn

Lookalikes: *A. excelsa* and *A. pantherina* lack red tints

The Blusher
Amanita rubescens

Named for the reddish staining which develops when the flesh is cut or bruised, and particularly where eaten by slugs. It is a common woodland toadstool, edible only after thorough cooking. It is poisonous raw, containing haemolytic compounds which cause the breaking up of red blood cells. It is best avoided anyway due to possible confusion with poisonous species such as *A. pantherina*.

AGARICACEAE

The Parasol
Macrolepiota procera

ID FACT FILE

Cap: 10–25 cm, at first ovoid, expanding to parasol shape, pale buff with brown, soft scales

Stem: Tall, up to 35 cm, slender, with brown snake-like pattern and movable, double, woolly ring

Gills: Free, soft, whitish. Spores white

Habitat: Fields, parks

Season: Summer–autumn

Lookalikes: Shaggy Parasol is less slender and without snake-like stem markings. *L. mastoidea* has simple ring

One of the best edible species, common throughout most of Europe and widely distributed in the N hemisphere. It may grow singly or in groups, usually in open parts of woods or in grass. Young fruitbodies are pestle-shaped, but soon expand, leaving a movable ring. The soft cap scales and snake-like markings on the stem are characteristic.

TRICHOLOMATACEAE

Plums-and-Custard
Tricholomopsis rutilans

ID FACT FILE

Cap: 4–10 cm, convex, densely covered with small reddish-purple scales on yellow ground

Stem: Cylindric, to 10 cm high, yellowish above, becoming reddish scaly downwards

Gills: Crowded, sinuate, yellowish. Spores creamy-white

Habitat: On rotten coniferous wood

Season: Late summer–late autumn

Lookalikes: None

The common name refers to the characteristic colour of the cap and stem. The species is distinctive and can scarcely be confused with any other. It is frequent throughout most of Europe on coniferous logs and stumps, especially of spruce. A related species, *T. decora*, is golden yellow; it is also on conifers but usually at higher altitudes, and is much less common.

TRICHOLOMATACEAE

Amethyst Deceiver
Laccaria amethystea

J	F	M	A	M	J
J	A	S	O	N	D

ID FACT FILE

Cap: 2–5 cm, convex becoming flattened, slightly scaly, deep violet

Stem: 5–8 cm high, cap colour, tough and fibrous

Gills: Adnate, thick and distant, violaceous, pow-dered white at maturity. Spores white

Habitat: Woods

Season: Summer–early winter

Lookalikes: None when fresh

Common, often in large groups on the ground in woods, occurring with both deciduous and coniferous trees. Easily recognised when in fresh condition by the violaceous colours throughout, but old and weathered specimens are much paler, more scurfy on the cap and may look very different. Quite a good edible species, also known as the Red Cabbage Fungus.

Abnormal, distorted specimens are quite commonly encountered.

TRICHOLOMATACEAE

The Deceiver

Laccaria laccata

ID FACT FILE

Cap: 2–5 cm, convex to flattened, red-brown or brick, drying paler, slightly scurfy, margin striate, often wavy

Stem: 5–8 cm high, sometimes compressed, cap colour, tough and fibrous

Gills: Pinkish, thick, distant, white-powdered from spores at maturity. Spores white

Habitat: Woods

Season: Summer –early winter

Lookalikes: *L. proxima* is more regular in form and with paler pink gills but distinguished with certainty only on spore characters. *L. bicolor* has a lilac stem base

A very common toadstool, growing singly or in groups in various kinds of woodland. It is very variable in appearance depending on age and conditions, hence the popular name. When dry, the cap can be much more scurfy and is pale to almost whitish although the gills usually retain their colour. The species is edible, but rather tough. The stem is usually rather twisted and compressed.

TRICHOLOMATACEAE

Clouded Agaric
Clitocybe nebularis

J	F	M	A	M	J
J	A	S	O	N	D

ID FACT FILE

CAP: 7–15 cm, convex to flattened, fleshy, greyish with whitish bloom. Flesh white

STEM: 7–12 × 2–3 cm, cylindric or slightly enlarged to base, smooth

GILLS: Whitish, decurrent, closely spaced. Spores pale pinkish

HABITAT: Mixed woods and heathland

SEASON: September–December

LOOKALIKES: *C. maxima* is larger and with central boss. *C. alexandri* more yellow-brown and much less common

A large, distinctive species sometimes growing in fairy rings in woods. The greyish cap has a characteristic whitish bloom composed of small tufts of hyphae. The odour of fresh specimens is said to be reminiscent of cottage cheese, and the species was once popularly known as the New Cheese Agaric. Although eaten by some, this fungus should be avoided; it is a stomach irritant to many. It is host occasionally to another toadstool, the rare *Volvariella surrecta*.

TRICHOLOMATACEAE

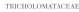

J	F	M	A	M	J
J	A	S	O	N	D

Buttercap
Collybia butyracea

Somewhat variable in colour but the damp, rather greasy feel to the cap and its dark centre are characteristic. The fibrous stem, enlarged towards the base, is also typical, and the colour contrast between the stem and gills is distinctive. A common woodland species, often continuing to be found after the first frosts. A pale form which occurs on richer soils is known as var. *asema*.

ID FACT FILE

Cap: 3–7 cm, moist, with central boss, dark brown, becoming paler on drying except at centre

Stem: Fibrous, finely grooved, thicker towards the base, cap colour

Gills: White, crowded, adnexed. Spores white

Habitat: Mixed woods, on acid soils

Season: Autumn–early winter

Lookalikes: *C. distorta* and *C. prolixa*, much rarer, have more slender stem and lack dark umbo

TRICHOLOMATACEAE

Velvet Shank

Flammulina velutipes

J	F	M	A	M	J
J	A	S	O	N	D

ID FACT FILE

CAP: 2–8 cm, convex, expanding, yellow-orange, smooth

STEM: Yellowish above, otherwise dark brown and distinctly velvety, usually curved

GILLS: Adnexed, pale yellow. Spores white

HABITAT: Decaying deciduous trees, especially elm

SEASON: Mostly early–mid-winter

LOOKALIKES: None; the dark, velvety stem and late season make it distinctive

Sometimes known as the Winter Toadstool, this appears late in the season, often after the first frosts, and can be found throughout the winter. It grows in tufts on various deciduous trees, but is most frequent on dead elms, causing a white rot of the sapwood. The dark brown velvety stem and yellow-orange cap are distinctive, and the species is edible.

TRICHOLOMATACEAE

J	F	M	A	M	J
J	A	S	O	N	D

Fairy Ring Champignon
Marasmius oreades

ID FACT FILE

CAP: 2–5 cm, creamy-buff when dry, convex, with broad central boss, smooth

STEM: To 8 cm high, pale buff, tough, dry

GILLS: Free to adnexed, rather distant, whitish to pale buff. Spores white

HABITAT: Grassland, in fairy rings

SEASON: Late spring–autumn

LOOKALIKES: Some *Collybia* species may be similar, but grow in woods

The commonest of the fairy-ring formers in grassland. Rings expand slowly over many years and can be recognised by the lush growth of grass at their inner edge. Toadstools appear in a zone just outside the lush grass. Extensive folklore originates from times when their existence was unexplained and attributed to the supernatural. It is said that the fairies danced in a ring on a summer evening, toadstools grew in their tracks and were used as seats by the tired elves. A good edible species, which also dries well, but take care not to gather with it poisonous *Clitocybe* species which can occur in similar places.

TRICHOLOMATACEAE

Lilac Bonnet-cap
Mycena pura

ID FACT FILE

CAP: 2–5 cm, conical, expanding to almost flat, lilac or pinkish, margin striate. Odour of radish

STEM: To 6 cm high, cap colour, smooth, base white-woolly

GILLS: Adnate or slightly decurrent, pale pinkish. Spores white

HABITAT: In woods

SEASON: Late summer–autumn

LOOKALIKES: *M. pelianthina* has dark gill edge

Common in both coniferous and deciduous woods throughout Europe, growing on the ground amongst leaf litter. Variable in colour; violaceous and white forms also occur, but all have a strong radish-like smell. This species contains toxins of the indole group which can cause hallucinations.

RUSSULACEAE

J	F	M	A	M	J
J	A	S	O	N	D

ID FACT FILE

Cap: 4–10 cm, convex to flattened, bright red or scarlet, smooth. Taste very hot

Stem: 3–8 × 1–1.5 cm, pure white, thicker to base, fragile

Gills: Adnate, white to pale cream, rather distant. Spores white

Habitat: Under pines

Season: Summer–late autumn

Lookalikes: *R. mairei* is very similar, but is found under beech

The Sickener
Russula emetica

A common species of pine woods, occurring throughout Europe. The scarlet cap, white stem and gills, and very hot, acrid taste are characteristic, but there are several red-capped *Russula* species and identification can be very difficult. This species, like others with an acrid taste, is inedible, being a stomach irritant, and causes sickness if eaten raw.

Blackening Russule

Russula nigricans

ID FACT FILE

CAP: 6–18 cm, convex, then depressed at centre, off-white becoming dark brown to blackish. Flesh reddening. Taste slowly hot

STEM: 3–8 × 1–3 cm, white to brownish or black

GILLS: Adnate, cream to straw, then blackening, thick, distant. Spores white

HABITAT: Woods

SEASON: Late summer–late autumn

LOOKALIKES: *R. densifolia* has much more crowded gills

A common woodland species, easily recognised by the large, robust fruitbodies which have thick, wide-spaced gills and flesh which reddens at first and becomes black with age. Old and rotting fruitbodies are commonly host to Pick-a-back Toadstool and the closely related *Asterophora parasitica*, and also to the small white *Microcollybia tuberosa* whose fruitbodies develop from small reddish-brown sclerotia, developed on the rotting flesh.

RUSSULACEAE

J	F	M	A	M	J
J	A	S	O	N	D

Common Yellow Russule
Russula ochroleuca

ID FACT FILE

Cap: 4–10 cm, convex to flattened, ochre-yellow, margin slightly striate

Stem: 4–7 × 1–2 cm, white then greying, often veined

Gills: Adnexed, whitish to cream. Spores white to pale cream

Habitat: Woods, usually with deciduous trees

Season: Late summer–late autumn

Lookalikes: *R. claroflava* is brighter yellow and has yellowish gills; *R. lutea* has egg-yellow gills

One of the commonest *Russula* species, occurring in various types of woodland. The yellow cap becomes paler with age and is slightly slimy in wet conditions. The grey colour in the stem develops with age, especially when wet, and is characteristic. The species has a mild or slightly hot taste and can be eaten if well cooked.

TRICHOLOMATACEAE

J	F	M	A	M	J
J	A	S	O	N	D

Wood Blewit
Lepista nuda

ID FACT FILE

Cap: 6–12 cm, convex, expanding, violaceous, fading to brown

Stem: Cap colour, swollen at base

Gills: Adnate, crowded, pale blue-violaceous. Spores pale pinkish

Habitat: Woods, parks, amongst leaf litter

Season: Late autumn–early winter

Lookalikes: Field Blewit has pale cap and gills. Some *Cortinarius* species have brown spores, and may be poisonous

This species typically appears late in the season, and is often found after the first frosts. It is fairly common, growing in grass and amongst leaf litter in parks and woods, and even in gardens. It is a good edible species, much sought after, in fact, as there are few others available late in the season, but it should not be eaten uncooked.

CORTINARIACEAE

Two-toned Wood-tuft
Kuehneromyces mutabilis

ID FACT FILE

Cap: 3–7 cm, convex or with central boss, orange-brown, drying pale ochraceous from the centre

Stem: To 8 cm high, with ring, dark brown scaly below ring, smooth and paler above

Gills: Adnate, pale cinnamon. Spores ochre-brown

Habitat: Stumps of deciduous trees

Season: Summer–autumn

Lookalikes: *G. marginata* on conifers and *G. autumnalis* on deciduous wood lack scaly stem

Common, growing in tufts on rotten stumps of a variety of deciduous trees. The cap is distinctly hygrophanous, drying markedly paler from the centre outwards and forming a darker marginal zone, hence the popular name. This is a good edible species, but must not be confused with the highly toxic Autumn Pixy-cap.

STROPHARIACEAE

Shaggy Scale-head
Pholiota squarrosa

| J | F | M | A | M | J |
| J | A | S | O | N | D |

ID FACT FILE

Cap: 4–11 cm, convex, expanding, tawny-yellow, dry, densely covered with recurved russet-brown scales

Stem: To 15 cm high, scaly below ring, tapered at base, cap colour

Gills: Adnate, crowded, pale yellowish, then rusty. Spores brown

Habitat: In clumps at base of trees

Season: Late summer–late autumn

Lookalikes: *P. squarrosoides*, has somewhat greasy cap and smaller spores; *P. aurivella* has sticky cap with adpressed scales and grows high up on trees

A large and distinctive species which grows in clumps, usually at the base of deciduous trees, especially beech and ash. It is a weak parasite, entering the tree through wounds. Recognised by the habit and the dry, densely scaly cap and stem. Slightly poisonous, often causing digestive upsets.

CORTINARIACEAE

J	F	M	A	M	J
J	A	S	O	N	D

Freckled Flame-cap

Gymnopilus penetrans

ID FACT FILE

Cap: 2–7 cm, convex to flattened, bright rusty-tawny, slightly fibrillose

Stem: 3–6 cm high, 4–7 mm thick, yellow, fibrous

Gills: Adnate, yellowish to tawny, soon rusty spotted, crowded. Spores yellow-brown

Habitat: On branches and wood, especially in pine woods

Season: Late summer–late autumn

Lookalikes: *G. sapineus* is more scaly and restricted to conifers

Found throughout Europe, and generally a common species, sometimes fruiting in abundance. It is readily recognised by the rusty colours and spotted gills. Specimens with a veil leaving an indistinct ring-zone on the stem are sometimes separated as a distinct species, *G. hybridus.*

AGARICACEAE

J	F	M	A	M	J
J	A	S	O	N	D

Horse Mushroom
Agaricus arvensis

ID FACT FILE

CAP: 10–15 cm, convex, whitish bruising yellowish, scales lacking

STEM: Up to 12 × 2.5 cm, white, cylindric, slightly thicker at base. Ring spreading, with cogwheel of soft, brownish scales beneath

GILLS: Whitish at first, dark chocolate-brown at maturity, crowded, free. Spores dark brown

HABITAT: Meadows, parks, in grassy places

SEASON: Late summer–autumn

LOOKALIKES: Yellow Stainer

This is a good, edible species with a pleasant smell. It is fairly common, especially in fields and meadows in autumn, sometimes growing in fairy rings. The large, domed, white cap is characteristic, as is the yellowing of the surface with age or on bruising. It must not be confused with the poisonous Yellow Stainer which smells unpleasant and also differs in having the stem base bright yellow when cut.

AGARICACEAE

J	F	M	A	M	J
J	A	S	O	N	D

Pavement Mushroom

Agaricus bitorquis

The popular name of this species is quite apt; it is well known for its habit of pushing up paving stones, providing a good example of the power generated by growing hyphae. It is a good edible species, recognised by the presence of two rings on the stem, the lower one sheathing the stem base rather like a volva. Take care not to confuse it with poisonous *Amanita* species.

ID FACT FILE

CAP: 5–10 cm, fleshy, convex to flattened, whitish, not scaly

STEM: 6–9 × 2–3 cm, whitish, with two rings

GILLS: Pinkish at first, becoming dull red-brown. Spores dark brown

HABITAT: Road-sides, pavements, wasteland

SEASON: Summer–autumn

LOOKALIKES: Other *Agaricus* species have a simple ring

COPRINACEAE

Common Ink-cap
Coprinus atramentarius

ID FACT FILE

CAP: 3–6 cm across, 3–7 cm high, oval or bell-shaped, expanding, greyish-white or pale fawn, slightly scaly at centre

STEM: Whitish, fragile, slightly thicker downwards, then rooting below ridge-like ring-zone

GILLS: Narrow, free, greyish, becoming black and deliquescing as the spores develop. Spores blackish-brown

HABITAT: On soil in woods and parks, growing from buried wood

SEASON: Early summer–late autumn

LOOKALIKES: *C. acuminatus* has narrower, more pointed cap and narrower spores, uncommon in woods

One of the most common of the ink-caps, usually growing in large tufts, and often found in man-made habitats, sometimes even pushing up through asphalted surfaces. It was once actually used for making ink, by boiling the mature, blackening caps. This species is edible, but must never be consumed with alcohol when it causes nausea, sweating and palpitations – symptoms identical to those of antabuse used in the treatment of alcoholism.

COPRINACEAE

J	F	M	A	M	J
J	A	S	O	N	D

Shaggy Ink-cap
Coprinus comatus

ID FACT FILE

Cap: 5–12 cm high, cylindric then narrowly conical, whitish, shaggy, brownish at the top. Flesh thin

Stem: Tall, hollow, white, with thin, loose ring below, and rooting base

Gills: Ascending, free, white at first, soon pinkish from the base, blackening and deliquescing. Spores blackish

Habitat: In grass, lawns, roadside verges etc.

Season: Summer–late autumn

Lookalikes: Unlikely to be confused with other species when in good condition. *C. sterquilinus* is smaller and on dung

Also known as Lawyer's Wig, this is a very common species, found in all kinds of grassy areas, roadsides and gardens. Readily recognised by the shaggy, narrowly conical cap which soon blackens from below and deliquesces into an inky fluid. It is a good edible species when young, while the gills remain white, but must be eaten very soon after gathering.

COPRINACEAE

| J | F | M | A | M | J |
| J | A | S | O | N | D |

Glistening Ink-cap
Coprinus micaceus

ID FACT FILE

Cap: 2–4 cm across, bell-shaped or bluntly conical, expanding, grooved, pale orange-brown, covered with tiny mica-like particles

Stem: Whitish, fragile, hollow, variable in length

Gills: Whitish at first, soon brown and finally blackish. Spores blackish-brown

Habitat: Woodlands, parks, etc., usually around old stumps

Season: Spring–late autumn

Lookalikes: *C. truncorum* differs only in microscopic characters, but is rare

A common species, growing in large clusters on and around old stumps and logs of deciduous trees. Named for the shining, mica-like particles which cover the cap in young fruit-bodies; it has also been known as the Glittering Toadstool. Unlike some ink-caps, this shows only slight deliquescence and does not produce an inky fluid. It is generally regarded as an edible species, but cannot be recommended.

STROPHARIACEAE

| J | F | M | A | M | J |
| J | A | S | O | N | D |

Sulphur Tuft
Hypholoma fasciculare

A common species which grows in dense tufts on rotten stumps and trunks of both deciduous and coniferous trees. The greenish gills of young fruitbodies are distinctive, as is the sulphur-yellow cap, though this is often covered eventually with a deposit of purple-brown spores from surrounding caps. The species is not poisonous but has a bitter taste and should not be eaten.

ID FACT FILE

CAP: 2–6 cm, pale sulphur yellow, bluntly conical, expanding, smooth, but often with scale-like remains of the veil at the margin

STEM: To 10 cm high, yellowish, tawny towards base, with slight ring-zone, trapping dark spores

GILLS: Adnate, crowded, greenish-yellow, darkening as spores develop. Spores purple-brown

HABITAT: Decaying stumps and logs

SEASON: Early summer–early winter

LOOKALIKES: *H. sublateritium* has brick-coloured cap and stem. *Pholiota alnicola* has brown spores

STROPHARIACEAE

J	F	M	A	M	J
J	A	S	O	N	D

Verdigris Toadstool
Stropharia aeruginosa

ID FACT FILE

CAP: 3–7 cm, convex or blunt-conical, blue-green fading to yellowish, slimy, with white-cottony veil scales especially near the margin. Flesh whitish

STEM: Cylindrical, 4–9 cm high, 5–9 mm thick, rather slimy, covered in white-cottony scales below the distinct ring

GILLS: Adnate or slightly decurrent, pale greyish to violaceous-grey. Spores purple-brown

HABITAT: Damp woodland

SEASON: Late summer–late autumn

LOOKALIKES: *S. cyanea*, common in damp grassy places, has a poorly developed ring

One of a group of closely related and attractive species with cap bright blue-green when fresh, but soon fading to yellowish. This attractive species, with a well-developed ring, is rather less common than some others of the group, but occurs in various kinds of woodland habitats, as well as in parks and grassy places. It is known to contain small quantities of psilocybin, and is likely to cause mild hallucinations if eaten.

BOLETACEAE

| J | F | M | A | M | J |
| J | A | S | O | N | D |

Cep
Boletus edulis

ID FACT FILE

Cap: 8–15 cm, hemispherical, often becoming flattened, brown, fleshy

Stem: 8–15 × 3–8 cm, stout, swollen below, whitish to buff, with distinct network of raised lines

Pores: Small, whitish to greenish-yellow, not staining when bruised. Spores yellow-olive

Habitat: Coniferous and mixed woods

Season: Summer–autumn

Lookalikes: *B. aereus*, with oak, has dark brown cap and stem. *B reticulatus*, with beech, has stem netted to base. *B. pinophilus*, under pines, has dark brown cap

Also known as the Penny Bun. The most sought after of the boletes, this is an excellent edible species which can be dried for winter use. Of commercial importance, it is the species that is commonly used in mushroom soups. Sometimes common in coniferous woods, while closely related forms, such as those referred to as *B. aereus* and *B. reticulatus*, grow in deciduous forests.

XEROCOMACEAE

J	F	M	A	M	J
J	A	S	O	N	D

Bay Bolete
Xerocomus badius

ID FACT FILE

CAP: 4–12 cm, bay brown, convex, smooth. Flesh whitish, blue tinge on cutting

STEM: 4–10 × 0.8–2.5 cm, streaky, brownish, paler than cap

TUBES: Yellow, pores readily bruising blue-green

HABITAT: Mixed woods

SEASON: Late summer–late autumn

LOOKALIKES: *B. spadiceus*, very rare, has network on stem

Common in coniferous and sometimes deciduous woods, and a good edible species with firm flesh which has a pleasant nutty taste. It is somewhat variable in colour, but the streaky, brownish stem and blue-bruising pores are diagnostic. This species is sometimes placed in the genus *Boletus*.

PAXILLACEAE

| J | F | M | A | M | J |
| J | A | S | O | N | D |

Brown Roll-rim
Paxillus involutus

ID FACT FILE

CAP: Brown, convex then depressed at centre, sticky when damp, margin velvety, inrolled

STEM: 4–7 × 2–3 cm, cap colour

GILLS: Decurrent, yellow-brown, bruising purple-brown. Spores white

HABITAT: Woods, with birch

SEASON: Summer–autumn

LOOKALIKES: *P. rubicundulus* stains rust-brown and grows under alder

Aptly named because of the strongly inrolled cap margin, remaining so even in old specimens. The margin is at first rather woolly and has a characteristic 'scalloped' pattern of fine branching ridges. The gills bruise dark brown and can be easily removed from the cap, as can the tubes in boletes, to which *Paxillus* is closely related. A poisonous species, the toxin is cumulative causing potentially fatal haemolysis.

PAXILLACEAE

| J | F | M | A | M | J |
| J | A | S | O | N | D |

False Chanterelle
Hygrophoropsis aurantiacus

ID FACT FILE

CAP: 3–7 cm across, flat to depressed, orange-yellow, paling with age, dry and finely downy, margin inrolled

STEM: Tapered and often curved, 2–5 cm high, yellowish to orange-brown

GILLS: Decurrent, crowded, forked, deep orange. Spores white

HABITAT: In pine woods and on heaths

SEASON: Late summer–late autumn

LOOKALIKES: Chanterelle has thicker gills and is egg-yellow

A common species in coniferous woodland and often mistaken for the true chanterelle which may grow in similar places. It differs especially in the thinner, crowded gills, and in colour. Although eaten by some people, this species may cause gastric upsets and is best avoided. A whitish form found in damp woods on acid soil is sometimes distinguished as a separate species, *H. pallida*.

whitish form *H. pallida* (below)

LENTINACEAE

| J | F | M | A | M | J |
| J | A | S | O | N | D |

Oyster Mushroom
Pleurotus ostreatus

ID FACT FILE

Cap: 5–15 cm across, convex to flattened, greyish-brown, often with blue or lilac tints, fading, smooth or slightly scaly, margin wavy when old

Stem: Lateral or very eccentric, short, sometimes lacking, whitish, base woolly

Gills: Decurrent, whitish. Spores pale lilac

Habitat: In tiers on stumps and trunks of various trees

Season: All year, but mainly autumn

Lookalikes: *P. pulmonarius* is whitish and fruits mainly in late summer

A well-known and sought-after edible fungus, increasingly grown commercially. The cap colour is very variable, and forms with strong blue tints (var. *columbinus*) are known, popularly called Peacock Fungus. Though perhaps most frequent on beech, it can occur on many trees, including conifers, and not uncommonly on worked timber. It causes a rapid decay, in the form of a white, flaky rot.

CANTHARELLACEAE

| J | F | M | A | M | J |
| J | A | S | O | N | D |

Chanterelle

Cantharellus cibarius

ID FACT FILE

CAP: 4–9 cm across, convex to flattened, soon depressed at centre, margin inrolled, wavy and often lobed, egg-yellow. Faint apricot-like odour

STEM: 4–7 cm high, tapered downwards, cap colour or paler, smooth, solid

GILLS: Deeply decurrent, rather thick, with blunt edges, shallow, vein-like, irregularly forked, cap colour. Spores ochraceous

HABITAT: Coniferous and deciduous woods, especially on sandy, acid soil

SEASON: Late summer–autumn

LOOKALIKES: *C. friesii* is smaller, in beech woods. False Chanterelle is more orange on cap, with thin gills

This is one of the best edible fungi. It is much sought after and often seen for sale in Continental markets. It has an excellent flavour but is rather tough and should be cooked slowly. There is a white form of this species, and another, with reddish-purple scales, is known as var. *amethysteus*. The Chanterelle is seemingly less common now in many areas, perhaps due to pollution from acid rain.

POLYPORACEAE

J	F	M	A	M	J
J	A	S	O	N	D

Dryad's Saddle
Polyporus squamosus

ID FACT FILE

CAP: 12–30 cm across, semi-circular, yellow-brown with dark brown, flat, concentric scales

STEM: Eccentric or lateral, short, blackish at base, apex with a network formed by decurrent pores

TUBES: Whitish, decurrent. Pores large, angular, edges often torn. Spores white

HABITAT: On stumps and trunks of dead or dying deciduous trees

SEASON: Early summer–autumn

LOOKALIKES:
P. tuberaster has rougher scales and grows from a large underground sclerotium. *P. mori* is much smaller and grows on gorse stems

A common and highly distinctive species, with a scaly, yellow-ochre cap – hence the other common name 'Scaly Polypore'. This species occurs especially on ash, elm and sycamore, on logs, stumps and even living trunks. It sometimes forms strange, stag-horn like structures if developed in the dark. It may also develop as a parasite, often entering the host where branches have been broken off, and causes a heart rot which can result in the trunk of the tree becoming hollowed out.

CORIOLACEAE

Many-zoned Polypore
Trametes versicolor

ID FACT FILE

FRUITBODY:
Bracket-like,
often in tiers,
3–8 cm across,
1–4 cm wide.
Upper surface
zoned in various
colours, velvety
at first, smooth
with age. Flesh
tough, thin, flexible, whitish.
Tubes short,
pores small,
whitish. Spores
white

HABITAT: On dead
stumps and logs
of various trees

SEASON: All year

LOOKALIKES:
T. pubescens is
whitish and lacks
coloured zones.
T. hirsutus is
more hairy, less
zoned and paler.
T. multicolor has
thicker flesh

One of the commonest of the bracket fungi, often forming large groups. It occurs on a range of hardwoods, including structural timber, and is occasionally found on conifers. The cap is always distinctly zoned, but very variable in colour. It does not change much in appearance on drying, and was once used as jewellery.

CORIOLACEAE

Chicken-of-the-Woods
Laetiporus sulphureus

ID FACT FILE

Fruitbody: Thick, irregular, with separate brackets arising from common base, 15–40 cm across. Upper surface yellow or orange at first with sulphur-yellow margin, fading to whitish with age. Flesh thick, whitish. Pores small, sulphur-yellow when young. Spores white

Habitat: On stumps and logs of various trees

Season: Early summer–autumn

Lookalikes: None

One of the most distinctive of the bracket fungi due to its bright yellow and orange colours. With age the colours fade to whitish, and the flesh becomes cheese-like and crumbly. It is parasitic, entering through wounds and causing a brown, cuboidal rot of the heart wood. However, infected trees may live for many years. Often eaten when young and fresh, but can cause allergic reactions.

CORIOLACEAE

J	F	M	A	M	J
J	A	S	O	N	D

Birch Polypore
Piptoporus betulinus

ID FACT FILE

FRUITBODY:
Bracket-like with narrowed attachment, 10–20 cm across, 3–8 cm thick, at first rounded, soon hoof-shaped, upper surface convex, pale brown, smooth, skin separable. Flesh firm, white. Pores small, whitish. Spores white

HABITAT: On trunks of birch

SEASON: All year

LOOKALIKES: None

A very common polypore, found on most dead and dying trunks of birch, sometimes many on a tree. It attacks both the heartwood and the sapwood, and eventually causes the top of the tree to fall out. The firm flesh of this species is ideal for stropping razors and was once used for this purpose; indeed, it has the alternative common name of Razor-strop Fungus.

J	F	M	A	M	J
J	A	S	O	N	D

Root Fomes
Heterobasidion annosum

ID FACT FILE

FRUITBODY:
Resupinate or
forming a bracket, up to 15 cm
across, 2–3 cm
thick, surface
uneven, brown to
orange-brown,
margin white.
Flesh firm, tough,
whitish to cream.
Pores small,
white. Spores
white

HABITAT: Base of
living trunks of
conifers, sometimes other trees

SEASON: All year

LOOKALIKES: None;
the hard flesh
and white margin
are distinctive

A serious parasite of conifers, forming brackets
at the trunk base and on the roots. This species
is especially damaging to plantation conifers
but also occurs in native forests, causing a serious white heart rot of the host trunk. Brackets
are perennial and may be found throughout
the year. The firm, hard flesh and white margin make the species distinctive.

GANODERMATACEAE

J	F	M	A	M	J
J	A	S	O	N	D

Artist's Fungus
Ganoderma adspersum

ID FACT FILE

FRUITBODY: A large, solid bracket, up to 60 cm across. Upper surface a dull grey-brown crust with concentric grooves. Flesh tough, rust-brown, to 10 cm thick. Tubes in annual layers; pores whitish, readily bruising brown. Spores brown

HABITAT: Common on beech, sometimes other trees, causing rot

SEASON: Perennial

LOOKALIKES: *G. applanatum* has thinner bracket with narrow margin, and white flecks in flesh

Very common, growing on various deciduous trees and causing a serious heart rot which is slow to develop but will eventually kill the tree. The pore surface is whitish but immediately bruises brown so that it is possible to draw or write on it, hence the common name. However, other species, notably *G. applanatum*, show a similar bruising.

Hairy Leather-bracket
Stereum hirsutum

ID FACT FILE

FRUITBODY: Bracket-like, 3–7 cm across, upper surface hairy, pale yellowish-grey, faintly zoned, margin wavy. Flesh thin, tough, leathery. Fertile surface smooth, bright yellow-orange, fading with age, not bleeding when cut. Spores white

HABITAT: Mostly on dead stumps and logs of deciduous trees

SEASON: All year

LOOKALIKES: Distinct, when fresh, in bright yellow-orange colours

A very common species, the commonest member of the genus throughout Europe. It occurs usually on fallen trunks and branches, rotting the sapwood, but is occasionally parasitic. It is variable in form, usually with well-developed brackets, but can sometimes be mostly resupinate, especially if growing on the underside of logs. The uncommon *Trenella aurantia* is parasitic on this species.

STEREACEAE

Silver-leaf Bracket
Chondrostereum purpureum

J	F	M	A	M	J
J	A	S	O	N	D

ID FACT FILE

FRUITBODY:
Resupinate, sometimes in extensive patches, with upturned, bracket-like margin, 2–4 cm across, often developed in tiers. Upper side whitish or greyish, hairy. Lower surface purplish, smooth. Spores white

HABITAT: On logs and stumps of deciduous trees, also parasitic, especially on rosaceous trees, causing silver-leaf disease

SEASON: Summer–autumn

LOOKALIKES: Some species of *Amylostereum*, which are on conifers

Occurs on a range of hardwood substrates, though frequently on birch. It is easy to recognise, at least in fresh condition, by the purple to violet-brown fertile surface. Though frequently a saprophyte, this species causes the important silver-leaf disease of plum trees and other rosaceous species. This affects the foliage, causing the leaf surface to develop a silvery sheen and trees may die in 2–3 years.

AURICULARIACEAE

| J | F | M | A | M | J |
| J | A | S | O | N | D |

Jew's Ear
Auricularia auricula-judae

The ear-like, gelatinous fruitbodies are edible and treated as a delicacy in some parts of the world. The species occurs on a range of trees but is most common on elder. The common name is a corruption of 'Judas's ear', referring to Judas Iscariot who supposedly hanged himself upon an elder tree after his betrayal of Christ to the Pharisees.

ID FACT FILE

FRUITBODY: Ear-shaped or irregular, 3–8 cm across, gelatinous, drying hard. Outer surface reddish-brown, finely hairy. Inner surface grey-brown, somewhat wrinkled. Spores white

HABITAT: On dead branches of deciduous trees, especially elder

SEASON: All year

LOOKALIKES: Some *Peziza* species, but these have brittle, non-gelatinous flesh

LYCOPERDACEAE

Common Puffball

Lycoperdon perlatum

ID FACT FILE

FRUITBODY:
4–9 cm high,
2.5–5 cm
across, rounded
above, with
stem-like base,
at first whitish,
soon yellowish-
brown; surface of
upper part
densely covered
with fragile, coni-
cal warts which
soon become
rubbed away to
leave a distinct
net-like pattern.
Opening by a
small pore.
Spore mass
olive-brown.
Sterile base
spongy

HABITAT: Woods

SEASON:
Summer–autumn

LOOKALIKES:
L. foetidum is
darker and has
spines in small
groups, and
united at their
tips

The commonest of the puffballs, usually
growing in groups and found in various kinds
of woodland. It is most easily recognised by the
conical spines which leave a net-like pattern
when rubbed away. The fruitbodies are edible
when young, whilst still white inside, but are
not recommended. In common with various
species of puffball, the spore mass of this fungus was once used as a styptic.

LYCOPERDACEAE

J	F	M	A	M	J
J	A	S	O	N	D

Giant Puffball
Calvatia gigantea

ID FACT FILE

FRUITBODY: Very large, commonly 20–50 cm across, rarely to 80 cm, subglobose or rather flattened, surface when young whitish, soft-leathery, sterile base lacking. At first white inside, spore mass at maturity olive-brown to dark-brown

HABITAT: Grass, parks, copses

SEASON: Summer–autumn

LOOKALIKES: Unmistakable when fresh and mature

One of the largest fungi in the world, unmistakable in appearance, well known and sought after as an excellent edible species. It is also probably the world's most prolific fungus, each fruitbody producing over 7,000,000,000,000 spores. Young fruitbodies whilst still white inside are delicious. It has had various other uses, for example as tinder, and in beekeeping, smouldering fruitbodies being placed beneath the hive to calm the bees. It is also a source of the anti-cancer drug calvacin.

GEASTRACEAE

J	F	M	A	M	J
J	A	S	O	N	D

Common Earth-star
Geastrum triplex

One of the most common and also one of the largest of the earth-stars, widely distributed in Europe and indeed throughout temperate and subtropical areas. Its large size and the presence of a distinct collar around the spore sac in most collections make it easy to recognise. The onion-shaped unexpanded fruitbodies are also distinctive, similar to but larger than those of the rare *G. lageniforme*.

ID FACT FILE

FRUITBODY: Onion-shaped at first, soon splitting into 5–7 rays which arch downwards to expose the fawn to pale buff spore sac. Expanded fruitbody 4–12 cm across, outer surface not encrusted. Spore sac unstalked, 2.5–3.5 cm across with single, fringed pore at the top. Inner layer commonly splitting around the spore-sac and turning upwards to form a distinct collar

HABITAT: On well-drained, calcareous soil, in woods or open areas

SEASON: Summer–late autumn

LOOKALIKES: *G. lageniforme* is smaller and rarer and lacks the collar around the spore sac

SCLERODERMATACEAE

J	F	M	A	M	J
J	A	S	O	N	D

Common Earthball
Scleroderma citrinum

ID FACT FILE

FRUITBODY:
5–12 cm across, rounded to rather depressed, lacking a stem-like base. Wall up to 5 mm thick, tough, surface yellowish to yellow-orange, usually with coarse scales. Spore mass greyish, with white veins, later purple-black, odour unpleasant

HABITAT: Woods and heaths, on acid soils

SEASON: Late summer–autumn

LOOKALIKES:
S. verrucosum and *S. areolatum* have smaller scales, and a stem-like base. *S. bovista* has a thin, almost smooth surface

This is the commonest of the earthballs, found throughout Europe and also in N America. Though used at times to adulterate truffles, this is a somewhat poisonous fungus if ingested in large quantities, and should never be eaten at any stage of its development. An uncommon bolete, *Boletus parasiticus*, can sometimes be found growing on the fruitbodies of the Common Earthball.

young fruitbodies (above)
old, dehisced fruitbody (below)

PHALLACEAE

Common Stinkhorn
Phallus impudicus

ID FACT FILE

FRUITBODY:
12–20 cm high,
arising from a
gelatinous egg-
like stage. Egg
3–6 cm long,
rounded, whitish,
with white, cord-
like mycelium.
Stem sponge-
like, hollow,
fragile, whitish,
with conical cap
attached only at
the top. Cap with
honeycomb-like
network of veins,
covered at first
by the slimy, foul
smelling, black-
ish-olive spore
mass

HABITAT: In soil
around stumps
in woods

SEASON:
Summer–autumn

LOOKALIKES:
P. hadriani has
pinkish eggs and
occurs in sand-
dunes

A common and well-known fungus often easily
located in woods by its strong, unpleasant
odour which attracts flies. Eggs of the
stinkhorn were once known as 'witches' eggs',
and the phallic shape of the expanded fruitbody has led to its use as an aphrodisiac.
Indeed, the species has a great deal of associated folklore, concerning either its smell or its
shape, and it is also popularly known as 'wood
witch'.

OTIDEACEAE

J	F	M	A	M	J
J	A	S	O	N	D

Orange Peel Fungus
Aleuria aurantia

ID FACT FILE

FRUITBODY:
2–10 cm across, cup-shaped, later irregular to flattened, inner surface bright orange to orange-yellow, paler with age. Outer surface paler, whitish, minutely downy. Flesh thin, brittle

HABITAT: Gregarious, on bare soil of paths, etc.

SEASON: Late summer–late autumn

LOOKALIKES:
Melastiza chateri is similar in colour but smaller, usually densely clustered and with brownish margin

A well-known fungus, usually easily recognised by the orange, rather brittle, cup-shaped fruitbodies. These grow in clusters, usually on damp, bare soil, often on paths. Most larger cup-fungi, though very varied in form and colour, have been referred at some time to the genus *Peziza*, now much restricted in application, and have been popularly known as elf-cups. This species is also known as Great Orange Elf-cup.

HELVELLACEAE

J	F	M	A	M	J
J	A	S	O	N	D

Common White Saddle
Helvella crispa

ID FACT FILE

Fruitbody:
Upright, 6–12 cm high, cap saddle-shaped, whitish to cream or pale fawn, underside smooth, stalk 3–5 cm high, 1–2 cm thick, deeply grooved and ribbed

Habitat: Woods, copses

Season: Late summer–autumn

Lookalikes: *H. lacunosa* is dark grey

One of the commonest of the saddle fungi, widely distributed in the N hemisphere, occurring especially on sandy soil in forests and at path edges. It is variable in size and shape, but always with a whitish cap and white stem which distinguish it from other species. The fruitbodies are edible if well cooked but somewhat poisonous if eaten raw.

XYLARIACEAE

Candle-snuff Fungus
Xylaria hypoxylon

ID FACT FILE

FRUITBODY:
Upright, to 7 cm high; when immature branched, antler-like, flattened, stalk black, hairy, upper part white and powdery. Mature, sexual, fruitbody black throughout, unbranched, pointed, upper fertile part thicker, cylindric, roughened

HABITAT: On dead stumps and logs

SEASON: All year

LOOKALIKES: Very distinctive. *X. carpophila* is much more slender and occurs on beech mast

A common and readily recognisable species occuring on various kinds of dead wood, and reported to have a mycelium which is luminous. The whitish, stag-horn-like fruitbodies of the conidial state are said to resemble snuffed-out candle wicks. These and the black sexual fruitbodies occur commonly together on the same stump or log. A related and similar but more slender species, *X. carpophila*, occurs on fallen beech mast.

XYLARIACEAE

| J | F | M | A | M | J |
| J | A | S | O | N | D |

Dead-man's Fingers
Xylaria polymorpha

ID FACT FILE

FRUITBODY:
3–8 cm high,
1–2 cm wide,
stalked, club-
shaped, often
irregular in form,
flattened and
contorted, black,
surface rough-
ened with minute
warts. Flesh firm,
tough, white.
Spores black

HABITAT: On old
stumps of
deciduous trees,
especially beech

SEASON: All year

LOOKALIKES: *X.
longipes* is more
cylindric and
grows on *Acer*

The fruitbodies, in the form of black, finger-like clubs, arise from rotten logs and stumps and it is not difficult to see how the popular name was derived. It is sometimes also known as Devil's Fingers. This is quite a common and widely distributed species, though because of its dull colour and its occurrence in shady wooded areas it often tends to be rather inconspicuous.

INSECTIVORES

| J | F | M | A | M | J |
| J | A | S | O | N | D |

Western Hedgehog

Erinaceus europaeus

ID FACT FILE

TOTAL LENGTH:
Up to 30cm

TAIL LENGTH:
2–4cm

FIELD CHARACTERS:
A short-legged,
pointed snout
insectivore,
covered with
spines

SIGNS:
Droppings are
soft, blackish,
usually with
insect remains

SOUND:
Very noisy,
making a wide
range of grunts,
snuffles and
snorts

SIMILAR SPECIES:
None in NW
Europe. Related
species occur
in E Europe and
S Europe

A familiar animal that is generally most abundant in suburban gardens, parks and woodlands. It is nocturnal and a frequent road casualty. Its main defence is to roll into a ball, presenting an attacker with a mass of spines, but these are no defence against motor cars. The hedgehog feeds on invertebrates, and also small mammals, birds' eggs, nestlings, frogs and carrion. It builds a nest, often in a compost heap, and gives birth to up to 5 blind, helpless young, which have only a few soft spines when born. Surprisingly agile, it can climb and swim well. Hibernates, occasionally emerging on milder days during winter to feed.

INSECTIVORES

Mole
Talpa europaea

ID FACT FILE

TOTAL LENGTH:
Up to 15cm

TAIL LENGTH:
Up to 4cm

FIELD CHARACTERS:
A small black
insectivore, with
a very short tail
and pink nose.
The eyes are tiny
and the forepaws
are very large

SIGNS:
Mole hills and
runs are clearly
visible in pasture
and lawns

SOUND:
Normally silent,
but squeaks can
be heard when
they fight

SIMILAR SPECIES:
Closely related
species occur in
S and SE Europe

A burrowing mammal that is rarely seen above ground. Its presence is usually obvious in meadows and gardens because of the mole hills or 'tumps' of earth pushed up from its tunnelling. The Mole's fur is short and velvety, and its forepaws are modified into powerful digging hands. It feeds mostly on earthworms and other invertebrates which it catches when patrolling its tunnels. The Mole is able to immobilise surplus prey by biting it, and feeding on it later. It gives birth to 3–4, naked and helpless young in an underground nest, usually inside a larger mound or 'fortress'.

INSECTIVORES

J	F	M	A	M	J
J	A	S	O	N	D

Common Shrew
Sorex araneus

ID FACT FILE

TOTAL LENGTH:
Up to 12.5cm

TAIL LENGTH:
Up to 4.7cm

FIELD CHARACTERS:
A small brown
insectivore, with
red tips to its
teeth; distinctly
tri-coloured –
greyish below,
brown on the
sides, and dark
brown on the
back

SIGNS:
No obvious
signs. Makes
nest under logs,
stones, planks
etc.

SOUND:
High-pitched
squeaks

SIMILAR SPECIES:
Pygmy Shrew;
other red-toothed
shrews

Rarely seen, the Common Shrew is more likely
to be heard squeaking in a hedgerow. It is active
by day and night, consuming large numbers
of insects, earthworms, molluscs and other invertebrates. Occasionally it will prey upon baby
mice or voles. It nests underground, or beneath
a fallen log, and produces up to 5 litters a year
with up to 7 young in each, which are born
naked and helpless. Short-lived, it rarely survives
more than a year. Although owls feed on shrews,
most cats when they have killed them do not eat
them, probably because of their musky odour.

BATS

J	F	M	A	M	J
J	A	S	O	N	D

Common Pipistrelle
Pipistrellus pipistrellus

ID FACT FILE

TOTAL LENGTH:
Up to 5cm

TAIL LENGTH:
Up to 3.5cm

FOREARM:
2.8–3.5cm

FIELD CHARACTERS:
Small, with a fluttery flight. Can be distinguished from its closest relatives by call

SOUND:
Ultrasonic calls 55–80kHz

SIMILAR SPECIES:
Pygmy Pipistrelle, which calls at a lower frequency

One of the smallest European bats, widespread and common in almost all habitats, and particularly common in the northern parts of its range. It is the species most commonly found in houses, and although its numbers have declined due to poisoning from timber treatment, some colonies are still large. They hibernate in clusters (up to 10,000 have been recorded in a single cluster) in caves, cellars, trees, churches and, more rarely, houses. Although most populations are fairly sedentary, some migrate and distances of over 700km have been recorded. In summer the females gather in nursery colonies of up to 1,000, and give birth to 1–2 naked, helpless young each year. Individuals have been recorded as living up to 16 years.

BATS

Serotine Bat
Eptesicus serotinus

ID FACT FILE

TOTAL LENGTH:
Up to 14cm

TAIL LENGTH:
Up to 6cm

FOREARM:
4.8–5.7cm

FIELD CHARACTERS:
Large and dark
brown above,
buff-brown below.
Looping flight

SOUND:
Ultrasonic
25–52kHz

SIMILAR SPECIES:
Northern Bat,
which has
pale-tipped fur;
noctules, which
are more
reddish-brown

A large bat, found mostly near open woodland, hedgerows and parks. The Serotine often travels a considerable distance (several kilometres) between its roost and feeding areas. Some populations also migrate between their winter and summer roosts, over distances of up to 330km. It has adapted to man-made environments, and is commonly found roosting in the roofs of houses. However, this has probably led to dramatic declines as over the past 40 years the roof timbers of many houses have been treated with persistent insecticides, many of which are toxic to bats. In summer females gather in nursery colonies of up to 100.

Common Rabbit
Oryctolagus cuniculus

J	F	M	A	M	J
J	A	S	O	N	D

ID FACT FILE

TOTAL LENGTH:
42–60cm

TAIL LENGTH:
Up to 7cm

FIELD CHARACTERS:
Compact shape, relatively long hind legs; long ears and a short white tail

SIGNS:
Characteristic round, pea-sized droppings

SOUND:
Generally silent, but scream violently when captured, and thump hind feet when alarmed

SIMILAR SPECIES:
Hares, which have black-tipped ears

A familiar and often common animal, very similar to the domestic varieties, which are descended from it. Rabbits live communally in warrens – extensive burrows – grazing on grass and other herbage in the immediate vicinity. In hard winters they also gnaw bark, and when they are abundant they may nest above ground, usually in dense vegetation. They have a characteristic hopping gait, and at the approach of danger bound to their burrows. The 3–8 young are born naked, blind and helpless in fur-lined nest. They emerge to feed above ground at about 3 weeks, and are sexually mature at around 5 months.

RABBITS AND HARES

| J | F | M | A | M | J |
| J | A | S | O | N | D |

European Brown Hare
Lepus europaeus

ID FACT FILE

TOTAL LENGTH:
Up to 70cm

TAIL LENGTH:
Up to 13cm

FIELD CHARACTERS:
Larger than
Rabbit, with
black-tipped ears

SIGNS:
Damage to
shoots, saplings
is the most
common. Also
well-defined
paths in
meadows

SOUND:
Generally silent;
screams when
injured or
captured; grunts
when boxing, but
this is rarely
heard

SIMILAR SPECIES:
Mountain Hare
which is smaller,
greyer or white,
with shorter ears

Hares are generally solitary or seen in small groups, but will occasionally gather in small herds. They occur in open woodlands and marshes, but are generally most abundant in agricultural and grassland habitats. They breed in most months of the year, but in early spring the 'boxing' behaviour is most apparent, and they also engage in chases. They do not burrow or make nests and young are born in the open. They are fully formed at birth, with their eyes open and are active within a few days. The large eyes of the hare enable it to have almost 360° vision. They feed on grasses, herbage, bark and shoots.

RODENTS

J	F	M	A	M	J
J	A	S	O	N	D

Red Squirrel

Sciurus vulgaris

ID FACT FILE

TOTAL LENGTH:
Up to 27cm

TAIL LENGTH:
Up to 20cm

FIELD CHARACTERS:
An unmistakable rodent, with bright colouring and a long bushy tail. In winter the ears are tufted

SIGNS:
The remains of pine cones, nuts and shoots often litter the forest floor

SOUND:
When disturbed, a scolding *chuck-chuck-chuck*

SIMILAR SPECIES:
Grey Squirrel, which has a distinctively patterned tail, is greyer, and never has ear tufts

Although widespread in woodlands and forests throughout most of Europe, the Red Squirrel has become extinct in most of England, and is declining in other parts of the British Isles. In competition with the introduced Grey Squirrel, the Red is found mostly in extensive coniferous forests, but the Red also suffers from virus epidemics. In colouring it is typically a bright orange-red but in many parts of its range is almost black. Builds nests (dreys) in tree holes and forks in branches, and has 2 litters a year of up to 7 naked, helpless young. Does not hibernate, but stays in its nest in severe weather.

J	F	M	A	M	J
J	A	S	O	N	D

Field Vole
Microtus agrestis

ID FACT FILE

TOTAL LENGTH:
Up to 14cm

TAIL LENGTH:
Up to c.5cm

FIELD CHARACTERS:
A blunt-headed
'mouse', greyish
brown, with a
short tail

SIGNS:
Runs

SOUND:
Generally silent,
but does squeak,
particularly when
populations are
very dense

SIMILAR SPECIES:
Other voles,
particularly the
Common Vole

Within its range, the Field Vole is often extremely abundant. It is found in a wide variety of habitats, but is particularly abundant in open grasslands, including pastures. It makes extensive burrows, just below the surface and feeds on grasses and roots. Populations are cyclic, building up, then crashing, and when they are in plentiful numbers they provide food for such predators as foxes, weasels, short-eared owls, and barn owls. The Field Vole can have up to 6 litters a year of 4–6 young. The young are born naked and helpless, but can be sexually mature in less than a month.

RODENTS

J	F	M	A	M	J
J	A	S	O	N	D

Bank Vole
Clethrionomys glareolus

ID FACT FILE

TOTAL LENGTH:
Up to 12cm

TAIL LENGTH:
Up to 6.5cm

FIELD CHARACTERS:
A blunt-headed
'mouse', with a
relatively long
tail and fairly
prominent ears;
it is reddish
brown

SIGNS:
Burrows, food
remains under
logs, sheet iron
etc.

SOUND:
Generally silent,
but squeaks
and chatters
occasionally

SIMILAR SPECIES:
Other voles,
which are either
greyer or have
shorter tails

Over most of its range, it is one of the most abundant small mammals, and is consequently a very important item of prey for many mammals and birds. It is most abundant in hedgerows and woodlands, where, although generally terrestrial living in burrows, it also climbs trees and bushes. It feeds mostly on berries, seeds, nuts and other vegetable matter, but also invertebrates including molluscs. Its presence can be identified by small groups of droppings, and nuts with circular holes and no clear tooth marks. Bank voles have up to 4 litters a year, of 3–6 young, born naked and helpless.

RODENTS

Yellow-necked Mouse
Apodemus flavicollis

ID FACT FILE

TOTAL LENGTH:
Up to 25cm

TAIL LENGTH:
Up to 13cm

FIELD CHARACTERS:
Sandy brown above, white below with a characteristic chest band

SIGNS:
Remains of food, often in old birds' nests or nest boxes

SOUND:
Squeaks and chattering

SIMILAR SPECIES:
Other wood mice, which mostly lack the chest band

A relatively large wood mouse which is widespread over much of Europe, although many of the populations are isolated from each other. This species often enters houses, particularly in late summer and autumn, and makes a considerable noise for such a small mammal when clambering around in attics. Its food consists of berries, nuts, seeds and fruit as well as invertebrates. Up to 3 litters a year are produced of up to 9 young which are born naked, blind and helpless. It sometimes nests in bird or bat boxes.

RODENTS

ID FACT FILE

TOTAL LENGTH:
Up to 45cm

TAIL LENGTH:
Up to 23cm

FIELD CHARACTERS:
The fur is
characterised
by a greasy
appearance,
and the tail is
long and scaly.
The colour is
variable, but
generally
brownish

SIGNS:
Burrows,
droppings,
toothmarks

SOUND:
Twitterings
and squeaks

SIMILAR SPECIES:
Black Rat, which
is more mouse-
like

Brown Rat

Rattus norvegicus

Almost ubiquitous in Europe, though less
common in Mediterranean regions, and in areas
away from human settlements. The Brown Rat
is a native of Asia, but its range spread gradually
from the east until, by the Middle Ages, it had
colonised most of Europe, reaching Britain at
the end of the 18th Century. It is a serious pest,
causing damage to property and spreading
disease. Highly adaptable, it occurs in most
habitats, usually associated with man, in towns,
suburbs and farms. Mainly nocturnal, feeding
on almost anything edible. Up to 5 litters of 7–8
young are produced each year.

RODENTS

Wood Mouse
Apodemus sylvaticus

ID FACT FILE

TOTAL LENGTH:
Up to 22cm

TAIL LENGTH:
Up to 11cm

FIELD CHARACTERS:
Small, long-tailed mouse, sandy coloured with large ears and eyes

SIGNS:
Feeding tables in old birds' nests

SOUND:
Generally silent

SIMILAR SPECIES:
Yellow-necked Mouse

The most widespread and common species of its group, the Wood Mouse is found over most of Europe. Its presence in Iceland, the Faröes and on most of the Scottish islands (where they are often larger than those on the mainland) is thought to be as a result of accidental introductions by Viking settlers from Scandinavia. It is an agile and often arboreal mouse and can be extremely abundant. It feeds on seeds, fruit and also small insects, molluscs and other invertebrates. Up to 4 litters are produced each year, consisting of 3–9 young.

RODENTS

Western House Mouse

Mus domesticus

J	F	M	A	M	J
J	A	S	O	N	D

ID FACT FILE

TOTAL LENGTH:
Up to 18cm

TAIL LENGTH:
Up to 9.5cm

FIELD CHARACTERS:
Small, grey mice,
that run (jump
less than wood
mice)

SIGNS:
Oval droppings,
spoiled stored
foods; gnawed
holes in skirtings
etc.

SOUND:
High-pitched
squeaks

SIMILAR SPECIES:
Other house
mice, particularly
Eastern House
Mouse,
M. musculus

Like the Brown and Black Rats, the Western
House Mouse is a colonist from Asia, and
is almost invariably found close to human
settlements, particularly in the north of its
range. Most activity takes place at night,
including feeding and drinking; during daylight
hours the House Mouse usually sleeps or
grooms. Until recently it was thought that
there was only a single species of house mouse
in Europe, but it is now known that there are
at least five, some of which can only be
identified in the laboratory. House mice have
up to 10 litters a year, consisting of up to 9
young, which can mature at 6 weeks.

CARNIVORES

J	F	M	A	M	J
J	A	S	O	N	D

Red Fox

Vulpes vulpes

ID FACT FILE

TOTAL LENGTH:
Up to 1.5m

TAIL LENGTH:
Up to 0.6m

FIELD CHARACTERS:
Reddish
coloration
and bushy
tail, usually
white-tipped.
The ears and
paws are usually
blackish

SIGNS:
Footprints,
odour; droppings
are characteristic

SOUND:
High-pitched
barks and yips
and a scream
in the breeding
season

SIMILAR SPECIES:
Dogs less
agile – foxes
are surprisingly
cat-like

One of the most widespread mammals in Europe, occurring in almost all habitats, from Arctic tundra to towns. It feeds on a wide variety of plant and animal matter, including voles, beetles, earthworms, berries, carrion, and in towns it raids garbage. It kills livestock, particularly free-ranging poultry, but the damage caused to lambs is probably exaggerated. The 4–6 cubs are born in a den (earth) in early spring and are blind and helpless. Dens are made in burrows, such as old badger setts or enlarged rabbit holes, or may be self-dug in banks; caves or drains may also be used. Dens are primarily used by breeding vixens. A fox's den is recognisable by the remains of its prey outside, and distinctive musky odour. This odour can also be recognised when a fox has crossed a path.

CARNIVORES

Stoat

Mustela erminea

J	F	M	A	M	J
J	A	S	O	N	D

ID FACT FILE

TOTAL LENGTH:
Up to 30cm

TAIL LENGTH:
Up to 14cm

FIELD CHARACTERS:
Long bodied,
reddish brown
with a black
tail tip

SIGNS:
None usually
obvious

SOUND:
Chitters
and squeaks

SIMILAR SPECIES:
Weasel, which is
smaller and
lacks black tail
tip

The Stoat is a small, long-bodied carnivore.
It is also known as the Ermine in winter when
it moults into white pelage in areas that are
habitually covered with snow; elsewhere white
individuals are only rarely seen. Widespread in
the British Isles the Stoat can be found in most
habitats with some cover, even marsh and
moorland. Dens are made in burrows, trees
and rock crevices, and are lined with the fur of
the Stoat's main prey, rodents. The black tail
tip is retained. The Stoat feeds on mammals
up to the size of Rabbits and young Hares
(considerably larger than themselves). It has
a single litter of 6–12 young that are born blind
and helpless. Stoats can be attracted by
squeaking.

CARNIVORES

Weasel
Mustela nivalis

ID FACT FILE

TOTAL LENGTH:
Up to 32cm, but
usually smaller

TAIL LENGTH:
Up to c.9cm

FIELD CHARACTERS:
Tiny, long-bodied,
reddish brown
above, white
below. Lacks
black tip to short
tail. Turns white
in winter in north

SIGNS:
Footprints

SOUND:
Chitters
and squeaks

SIMILAR SPECIES:
Stoat which has
a black tip to the
tail

One of the world's smallest carnivores, the
Weasel is like a diminutive Stoat. It preys
on mammals considerably larger than itself
but feeds mostly on mice and voles, as it
is small enough to follow them into their
burrows. The Weasel occurs in almost all
habitats. Size varies considerably, with
the largest individuals occurring in the
Mediterranean area, and the smallest in the
far north. One or 2 litters a year are produced
of 4–6 young, occasionally as many as 12.

CARNIVORES

| J | F | M | A | M | J |
| J | A | S | O | N | D |

ID FACT FILE

TOTAL LENGTH:
Up to 65cm

TAIL LENGTH:
Up to 20cm

FIELD CHARACTERS:
Small, ferret-like
carnivore

SIGNS:
Droppings (which
have unpleasant,
pungent smell),
food remains

SOUND:
Growls, chitters,
snarls and hisses

SIMILAR SPECIES:
Other polecats;
Stoat which is
smaller and red-
dish

Polecat
Mustela putorius

Widespread over most of Europe, the Polecat was exterminated from most of Great Britain by the end of the 19th Century, with only a small population surviving in Wales. This has expanded and its range is now spreading. Polecats are closely related to Domestic Ferrets, which often escape and survive in the wild. The Polecat preys on small mammals and birds, and often raids poultry (the name is derived from *poule-chat*, 'chicken cat'). It has a strong odour – hence the alternative name of Foulmart. The Polecat has a single litter each year of 2–12 young (kits), which are born blind and helpless.

CARNIVORES

J	F	M	A	M	J
J	A	S	O	N	D

Badger
Meles meles

ID FACT FILE

TOTAL LENGTH:
Up to 1m

TAIL LENGTH:
Up to 20cm

FIELD CHARACTERS:
A large grey
carnivore with
distinctive black
and white head

SIGNS:
Setts, latrine
pits, footprints,
scratching, tufts
of hair on barbed
wire

SOUND:
Rather noisy
with a range of
snuffling, grunts,
barks etc.

SIMILAR SPECIES:
Not likely to
be confused
with any other
species

Over much of Europe, the Badger has been persecuted, despite the fact that its main prey is earthworms and other relatively small animals; it will occasionally take nestling rabbits and similar prey. Badgers are most commonly found in areas with a mixture of grassland and woodland, and especially in deciduous woodland with clearings. In the UK the Badger is persecuted because it becomes infected with bovine TB. Badgers live in burrows (setts) which are often very extensive and can be occupied for hundreds of years. A badger will dig latrine pits for its droppings, and change the grasses and leaves used for bedding regularly. A litter of 1–5 blind and helpless cubs is born in the sett in late winter or early spring. Usually strictly nocturnal, in areas where it is not persecuted it may occasionally forage during the day.

Common Seal
Phoca vitulina

ID FACT FILE

TOTAL LENGTH:
Up to 2m

TAIL LENGTH:
Absent

FIELD CHARACTERS:
Has a rather domed, dog-like head, and when hauled out on land a rather plump short body

SIGNS:
Usually none

SOUND:
Short barks and grunts

SIMILAR SPECIES:
Grey Seal, which is larger

A relatively small seal once common around the North Sea but now virtually confined to the more remote and less disturbed areas. The Common Seal hauls out on beaches and sand bars in groups, which can number 500 or more. It breeds in small colonies, in summer, and the pups, which are born with the adult fur, can swim soon after birth. The breeding sites are often on tidal sand bars, as well as rocky coasts and beaches. Within hours of birth, the young can dive, and some are weaned at two weeks. They feed on crustaceans, molluscs and fish. Most remain in the area where they were born, but some have been found up to 300km from their birth place.

HOOFED MAMMALS

J	F	M	A	M	J
J	A	S	O	N	D

Fallow Deer

Cervus dama

ID FACT FILE

TOTAL LENGTH:
Up to 2.3m

TAIL LENGTH:
Up to 20cm

FIELD CHARACTERS:
Usually dappled;
males have
palmate antlers

SIGNS:
Droppings and
hoof prints

SOUND:
Male has barking
grunt during the
rut

SIMILAR SPECIES:
Sika Deer; White-
tailed Deer

The Fallow Deer is mostly seen in parks and semi-wild conditions, and is often partly domesticated. It originally occurred around the Mediterranean region, but has been introduced widely throughout Europe as far north as the British Isles and Scandinavia. It occurs in a range of colour forms, including blackish (melanistic) and white (albino). The normal wild colouring is buff-brown above with white spots in summer, darker and uniform brownish in winter. The male's antlers are distinctive, with a flattened palmate shape. They are gregarious and have 1–2 heavily spotted fawns each year.

HOOFED MAMMALS

J	F	M	A	M	J
J	A	S	O	N	D

Red Deer
Cervus elaphus

ID FACT FILE

TOTAL LENGTH:
Up to about 1.5m

TAIL LENGTH:
Up to about 35cm

FIELD CHARACTERS:
A large, reddish brown deer with a buff rump. The males carry branching antlers

SIGNS:
Droppings, 'frayed' trees, footprints

SOUND:
Calves and females bleat, and males roar

SIMILAR SPECIES:
Most other deer are smaller

One of Europe's largest land mammals, the Red Deer can be surprisingly elusive and difficult to observe in dense woodland. It occurs in a wide range of habitats, including forests, swamps and mountain moorlands. It is active throughout the day and night, but more nocturnal in areas where it is hunted. Red Deer live in segregated herds, except during the rut. Males defend their territories locking antlers with rival males and uttering loud roars. The single fawn (occasionally twins) is heavily spotted. Hunted throughout its range, numbers are nevertheless increasing in most areas. In Scotland dense populations are limiting the regeneration of trees and other vegetation.

| J | F | M | A | M | J |
| J | A | S | O | N | D |

Roe Deer
Capreolus capreolus

ID FACT FILE

TOTAL LENGTH:
Up to 1.3m

TAIL LENGTH:
Up to 4cm

FIELD CHARACTERS:
Small, with short
spiky antlers.
Whitish or pale
buff rump patch

SIGNS:
Extensive
'fraying' of trees
in its territory;
droppings, tracks

SOUND:
Both sexes bark
in alarm, and
during the rut
have a rasping
call

SIMILAR SPECIES:
Muntjac; Chinese
Water Deer

A small deer, with an almost vestigial tail,
widespread over most of Europe. Although
generally shy, in areas where it is not hunted
it may become diurnal, and be encountered
feeding along edges of woodland and even in
gardens. It occurs mostly in woodland and
woodland edge habitats, often emerging to feed
on crops. Otherwise feeds on grasses, brambles,
rose, some conifers and young leaves; in
autumn takes advantage of fungi, beechmast
and acorns. Both sexes are territorial and
solitary, and the young are born brownish-black
and heavily spotted along back and flanks. The
female usually has twins, sometimes a single
young, and more rarely triplets.

SEALS, WHALES AND DOLPHINS

| J | F | M | A | M | J |
| A | S | O | N | D |

Common Dolphin
Delphinus delphus

ID FACT FILE

TOTAL LENGTH:
Up to 2.6m

TAIL LENGTH:
Present as flukes

FIELD CHARACTERS:
Long beak and colourful markings

SIGNS:
Rides bow waves of ships

SOUND:
Echolocating clicks

SIMILAR SPECIES:
Other dolphins

A small, very beautifully marked dolphin – the markings soon fading after death. It is one of the most widespread dolphins in European waters, and is usually seen in schools swimming fast and close to the surface, only diving for a few mins, feeding on fish and squid. It often leaps completely clear of the water and also rides the bow waves of large whales as well as ships. In the past schools often joined together and could number several thousand, but the Common Dolphin has been hunted (particularly in the Black Sea) and has also suffered heavy mortality in fishing nets.

SEALS, WHALES AND DOLPHINS

| J | F | M | A | M | J |
| J | A | S | O | N | D |

Bottle-nosed Dolphin

Tursiops truncatus

ID FACT FILE

TOTAL LENGTH:
Up to 4m

TAIL LENGTH:
Present as flukes

FIELD CHARACTERS:
Grey, with a
short beak and
no distinct
markings

SIGNS:
Often attracted
to boats

SOUND:
Wide range of
squeaks and
chirps

SIMILAR SPECIES:
Other dolphins
and porpoises

Possibly the best know dolphin, this is
one of the species most commonly seen in
dolphinaria and in films. It usually occurs
in small schools, and is often seen riding the
bow waves of ships. Occasionally it enters
rivers, and is frequently seen close to the
shore. It occurs throughout European waters,
most commonly in the Mediterranean and
Atlantic. It has a single young and feeds on fish
and crustaceans. Sometimes it follows fishing
boats to catch the fish the boats disturb. Until
comparatively recently huge numbers were
killed in the Black Sea.

J	F	M	A	M	J
J	A	S	O	N	D

Killer Whale

Orcinus orca

ID FACT FILE

TOTAL LENGTH:
Up to 9.5m

TAIL LENGTH:
Present as flukes

FIELD CHARACTERS:
Large, upright
dorsal fin (larger
than most other
whales)

SIGNS:
Young individuals
are often curious
about boats

SOUND:
Uses clicks and
other calls for
navigation

SIMILAR SPECIES:
Pilot whales

Often referred to as the wolf of the sea, this
is a good description as it operates in packs and
is capable of tackling prey much larger than
itself. The Killer Whale can kill large whales
and seals, as well as smaller marine mammals,
birds and fish. It is very playful and often
'plays' with its smaller prey, tossing it into the
air and chasing it before finally killing it.
'Packs' of Killer Whales will harry larger
whales eventually cutting them to pieces. It has
one of the widest known ranges of any
mammal, occurring in all oceans, from the
tropics to the polar regions.

SEALS, WHALES AND DOLPHINS

J	F	M	A	M	J
J	A	S	O	N	D

Fin Whale
Balaenoptera physalus

ID FACT FILE

TOTAL LENGTH:
Up to 25m or
more

TAIL LENGTH:
Present as
a fluke

FIELD CHARACTERS:
Large size, with
a relatively large
dorsal fin

SIGNS:
'Blows' up to
6m high

SOUND:
Low frequency
calls

SIMILAR SPECIES:
Blue Whale

One of the largest whales, weighing up to
nearly 70 tonnes, it is nearly as large as the
Blue Whale. It was once widespread and
abundant in European waters, but was hunted
to the verge of extinction by the early part of
the 20th Century. It is unusual since it has
asymmetrical colouring, with a pale marking
only on the right hand side of the lower jaw.
Despite its huge size it feeds mostly on small
fish and plankton which it filters from the sea
using the baleen plates in its mouth. The Fin
Whale matures at about 6–7 years and has a
single young every other year.

J	F	M	A	M	J
J	A	S	O	N	D

Blue Whale
Balaenoptera musculus

ID FACT FILE

TOTAL LENGTH:
Up to 30m

TAIL LENGTH:
Present as flukes

FIELD CHARACTERS:
Huge size, mottled bluish colouring and small dorsal fin

SIGNS:
'Blow' of up to 12m at surface

SOUND:
Low frequency

SIMILAR SPECIES:
Fin Whale

The largest mammal that has ever lived, it is believed that the male can grow to over 30m and weigh more than 175 tonnes. The Blue Whale is a very fast swimmer, reaching speeds of 30 knots per hour. Like other baleen whales it feeds on small animal prey particularly plankton (krill). Females do not reach maturity until 5 years and give birth to a single young every 3 years. Once widespread in all the oceans, by the middle of the 20th Century it had been hunted to the brink of extinction. Despite protection most populations are only recovering very slowly. It is most likely to be seen in the Atlantic, particularly off NW Scotland and Ireland.

SEALS, WHALES AND DOLPHINS

J	F	M	A	M	J
J	A	S	O	N	D

Minke Whale

Balaenoptera acutorostrata

ID FACT FILE

TOTAL LENGTH:
Up to 9.2m

TAIL LENGTH:
Present as flukes

FIELD CHARACTERS:
A small rorqual,
with white on
flippers

SIGNS:
'Blow' is
relatively
indistinct

SOUND:
A wide range
of sounds have
been recorded
ranging from
4–200kHz

SIMILAR SPECIES:
All other rorquals
are larger

Sometimes known as the Lesser Rorqual, this is the smallest of the rorqual whales, and also slightly less streamlined. In addition to its relatively small size, it can be distinguished from all other rorquals by white bands on the flippers. Minke Whales usually live in small groups, but sometimes gather in schools of several hundred, feeding primarily on fish. It can be remarkably agile, breaching and jumping clear of the water, and coming down headfirst, often crashing with a splash. The Minke Whale is frequently attacked by the Killer Whale. It is the most abundant of the rorqual whales and is still hunted in several parts of the world, including off Norway.

J	F	M	A	M	J
J	A	S	O	N	D

A Centipede
Lithobius forficatus

ID FACT FILE

Size:
Body length
3–4 cm

Description:
A long, thin,
flattened animal,
reddish brown in
colour, with 15
pairs of legs at
one pair per seg-
ment. It has long
antennae

Food:
Invertebrates

Lookalikes:
L. variegatus
is very similar,
with minor
differences

Centipedes are mobile predatory arthropods, with just one leg per segment, compared to the two per segment of millipedes. This is one of the commonest species, being found in a wide variety of habitats and quite often coming into houses and sheds at night in search of prey. It is an aggressive predator, hunting down almost any suitable-sized creature including slugs, and even other centipedes. They are nocturnal, hiding under logs and stones by day. It is common and widespread except in the driest or coldest areas. Similar species *L. variegatus* has a flattened, orange-brown body, about the same size as *L. forficatus*, and is often found in gardens as well as hedgerows and woodlands. It is a nocturnal hunter, hiding under stones and in compost heaps during the day.

WOODLICE, ISOPODA

J	F	M	A	M	J
J	A	S	O	N	D

Woodlouse
Porcellio scaber

ID FACT FILE

SIZE:
Body length
16–19 mm

DESCRIPTION:
A greyish-brown,
flattened animal,
paler around the
edges. There is
usually a pale
line of dots down
the centre

FOOD:
All sorts of scav-
enged material

LOOKALIKES:
There are several
similar species

This is one of the largest and commonest of woodlice, of which there are a number of similar species. Woodlice suffer from not being able to prevent themselves drying out, so they spend most of their time out of the sunlight, hidden under stones or logs, and foraging at night. The females carry the eggs or young in a special pouch under the body, keeping them humid. This species is common in sheltered humid conditions, and is widespread throughout Britain and much of Europe. It is often found with *Oniscus asellus*, another common woodlouse, under logs or in compost heaps. *O. asellus* has a smooth outline, with no obvious junctions between the different sections of the body, and is shiny grey with pale blotches at the sides.

TICKS AND MITES, ACARI

J	F	M	A	M	J
J	A	S	O	N	D

Sheep Tick
Ixodes ricinus

ID FACT FILE

Size:
Body length
3–4 mm

Description:
A greyish-brown
oval creature,
with six legs as a
young larva, or
eight reddish
legs when ma-
ture. When fully
fed, they resem-
ble a small red-
dish bean

Food:
Mammalian
blood

Lookalikes:
There are several
rather similar
species

Also known as the Castor Bean, because of its
shape when fully swollen, this is the
commonest and most economically significant
of the ticks. They have a strange life-cycle,
taking about three years and involving high
mortality. The young larva is six-legged, and
it must find a mammalian host to take blood,
after which it moults into an eight-legged
nymph. The following year, this has to find
another mammal before it can become
mature. The female adults then feed again
before mating and laying eggs. They carry
various diseases, and can bite man and dogs.
Ticks can attach themselves to you or your
clothing when walking through long grass or
bracken, so check carefully for them after
walking, especially in early summer. They
are widespread and common almost
throughout.

J	F	M	A	M	J
J	A	S	O	N	D

Harvestman
Phalangium opilio

ID FACT FILE

SIZE:
Body length
5–6 mm

DESCRIPTION:
Greyish above
with paler mark-
ings, and white
below. Males
have long for-
ward-curved jaws

FOOD:
Invertebrates and
carrion

LOOKALIKES:
*Leiobunum rotun-
dum* is similar

Harvestmen are a small group, resembling spiders, but with a one-piece body. Most have long thin legs. This is one of the commonest of harvestmen, occurring almost everywhere, and frequently seen because it is less nocturnal than most other harvestmen. It has long legs and seems to float over vegetation and other surfaces with ease. The females have long ovipositors, with which the eggs are laid deep into the ground. They occur in a wide variety of habitats including gardens, woodland clearings, scrub and rough grassland. It is widespread throughout Britain and most of Europe.

SPIDERS, ARANEAE

Garden Spider
Araneus diadematus

J	F	M	A	M	J
J	A	S	O	N	D

ID FACT FILE

SIZE:
Body length
10–13 mm

DESCRIPTION:
Predominantly
brown or orange-
brown, with a pat-
tern of white dots
in an approxi-
mate cross on
the abdomen.
The legs are
bristly and band-
ed brown and
white

FOOD:
Insects

LOOKALIKES:
Other orb-web spi-
ders are rather
similar, including
A. quadratus, but
none has the
white cross

Garden spiders are the best known of all the
orb-web spiders, and most people with a
garden will have come across them. The orb-
web spiders build large roughly circular webs
with radiating spokes, usually well off the
ground on bushes, or between the branches of
trees, and sit at the centre waiting for prey.
When an insect flies or jumps into the web, the
spider rushes to it and quickly wraps it in silk.
Eggs are enclosed in silk bags where they
remain over the winter; hatchlings emerge the
following spring. If disturbed, the spider
retreats to a hidden area at the edge of the
web. Very common and widespread in gardens,
woods and scrub, throughout Britain and
Europe, becoming most noticeable in autumn.

SPIDERS, ARANEAE

J	F	M	A	M	J
J	A	S	O	N	D

Hunting Spider
Marpissa muscosa

ID FACT FILE

SIZE:
Body length
7–12 mm

DESCRIPTION:
Dark brown-
black, furry, flat-
tened body.

FOOD:
Insects

LOOKALIKES:
M. radiata is
similar in shape,
with a yellow-
brown midline

These furry hunting spiders have extremely
good eyesight. With a square-fronted carapace
and roud large front-facing eyes, they stalk
their prey and then leap onto it. They have an
elongated abdomen and the first leg is darker
and thicker than the rest. Males use this leg in
a courtship display, raising it into the air and
moving from side to side, before approaching
to touch the female. Reaching maturity in
spring and summer, they can be found on the
bark and lichen of trees and fences. *M. radiata*
(see LOOKALIKES) is found in marshy areas.

EARWIGS, DERMAPTERA

| J | F | M | A | M | J |
| J | A | S | O | N | D |

Common Earwig
Forficula auricularia

ID FACT FILE

SIZE:
Body length
10–13 mm

DESCRIPTION:
A shiny brown
cylindrical insect.
Although they
appear to be
virtually wingless,
the hindwings
are actually partly
concealed under
the modified
front wings.
Males have
pincer-like
claspers

FOOD:
Various types of
plant material

A familiar and abundant insect, with a wealth of folk-lore and tales concerning its life history. They are much the commonest earwig in Europe, and the only species seen regularly in Britain and many other areas. Mainly nocturnal, hides during the day under bark or flower pots, emerging at night to feed. Feeds on plant material, living and dead, occasionally insects. Females look after the young and family groups may be discovered sheltering under objects in gardens. They occur in a wide variety of habitats, wherever there is food and shelter, and are often found in gardens and parks. The young are like miniature adults, and the female guards them until they disperse. There are several similar species, but these are much less common.

Common Field Grasshopper

Chorthippus brunneus

J	F	M	A	M	J
J	A	S	O	N	D

ID FACT FILE

SIZE:
Body length up to 18 mm (male), 25 mm (female)

DESCRIPTION:
A variable but generally brownish insect, with wings stretching well beyond the abdomen tip. The upper side of the tip of the abdomen is usually reddish, especially in males

FOOD:
Low vegetation, mainly grasses

LOOKALIKES:
The Bow-winged Grasshopper *C. biguttulus* has a markedly curved front edge to the forewing, especially in males. Widespread in north Europe, but absent from Britain

A medium-sized grasshopper, with noticeably long wings, and it flies more readily than most grasshoppers. It occurs in warm dry areas with short turf, such as chalk or limestone grassland, and not infrequently on mown lawns. The call is a short harsh buzz, lasting less than half a second, and repeated at 2–3 second intervals, with other males replying if they are nearby. It occurs throughout much of the central part of Europe, from the Pyrenees to Finland, and it is widespread and common in Britain.

Great Green Bush-cricket

Tettigonia viridissima

J	F	M	A	M	J
J	A	S	O	N	D

ID FACT FILE

SIZE:
Body length up to 42 mm, with the wings considerably longer

DESCRIPTION:
Green, except for a brown stripe down the back. The long robust wings; ovipositor 2cm long.

FOOD:
Omnivorous, mainly invertebrates

This impressive insect is the largest bush-cricket in north Europe, and one of the largest in Europe. You can often sense it moving through vegetation by the bending branches, before you actually see it! Despite its bulk, it is surprisingly hard to see. The call is characteristic, like a freewheeling bicycle, audible from a considerable distance, and emitted from around midday to the middle of the night. It occurs in a variety of bushy habitats, though in Britain it is largely coastal. Found only in Europe, the Upland Green Bush-cricket *T. cantans* is slightly smaller with wings only reaching just beyond the abdomen.

J	F	M	A	M	J
J	A	S	O	N	D

Blue-tailed Damselfly

Ischnura elegans

ID FACT FILE

SIZE:
Abdomen length
22–29 mm

DESCRIPTION:
A predominantly
black or dark
blue species,
with blue stripes
on the thorax,
and a blue seg-
ment at the tip of
the abdomen.
Females are
rather variable in
colour and pat-
tern

FOOD:
Small insects

LOOKALIKES:
Scarce Blue-
tailed Damselfly
is smaller and
more delicate,
and is rare in
Britain

A distinctive medium-sized damselfly, easily
picked out by its combination of almost black
body except for a single segment of bright blue
near the tip. It is a common and widespread
species, occurring around most types of water
bodies, and able to withstand some degree of
pollution and salinity. They rarely stray far
from the marginal vegetation of their home
water body, and may be quite inactive in dull
weather. It occurs almost throughout Europe,
except for the far north and Spain. In the
British Isles, the Blue-tailed Damselfly occurs
in much of England, Wales and Ireland. It is
less common in Scotland, found mainly in the
south of the region.

J	F	M	A	M	J
J	A	S	O	N	D

Four-spotted Chaser
Libellula quadrimaculata

ID FACT FILE

Size:
Abdomen length
27–32 mm

Description:
A distinctive
species, yellow-
ish-brown overall
with a bold black
tip to the ab-
domen; the
wings have yel-
lowish veins, and
four black spots
in addition to the
basal black trian-
gle that other
chasers have.
Males and fe-
males are very
similar

Food:
Small or medium
insects, caught
on the wing

Lookalikes:
Broad-bodied
Chaser (female)

A medium-sized dragonfly, with a fast darting
flight. The males are very active, aggressively
defending their territory against other males
throughout the day, but regularly returning to
a perch by the water. They prefer still water,
usually acid to neutral, such as lakes, ponds
and bog pools, where they may become
extremely abundant in favourable conditions.
In certain years, large quantities may migrate
westwards from eastern Europe. It occurs
throughout Europe except in the extreme
south. The Broad-bodied Chaser (*L. depressa*)
is another common species of dragonfly.
Smaller than the Four-spotted Chaser, it has a
broad body and strong flight which gives the
impression of a larger insect. Males are bright
blue, but females are yellowish-brown and can
be distinguished from *L. quadrimaculata* by its
dark wings with a brown triangle at the base.

J	F	M	A	M	J
J	A	S	O	N	D

Emperor Dragonfly
Anax imperator

ID FACT FILE

SIZE:
Abdomen length 50–60 mm, with males being the larger

DESCRIPTION:
Males are pre-dominantly bright turquoise blue, with a black stripe along the abdomen, and green on the head and thorax. Females are greenish-yellow with brown markings

FOOD:
Medium to large insects, including butterflies

LOOKALIKES:
Lesser Emperor is similar

The Emperor Dragonfly is one of the largest and most impressive insects in Europe. They are difficult insects to approach, as they are constantly on the wing, most often out over water, and when they do settle, it is often up in a tree. It is a fast-flying insect, usually keeping still over water. Females are rather less active. Their main habitat is larger ponds and lakes, or slow-moving water such as canals, and it will survive in brackish or slightly polluted waters. Often breeds in small ponds and ditches. It is very widespread and quite common through-out southern and central Europe, as far north as central England and southernmost Scandinavia.

J	F	M	A	M	J
J	A	S	O	N	D

hibernates

ID FACT FILE

SIZE:
Body length
10–12 mm

DESCRIPTION:
A long, narrow
bug though
quite robust in
appearance. It is
generally brown-
ish or brown-grey
in appearance,
with a noticeably
long, curved beak
with which it
attacks dead and
stranded insects

FOOD:
Scavenges for
insect remains
on the surface,
including any-
thing that falls
in the water

LOOKALIKES:
There are several
similar species,
e.g. *Aquarius
najas* which is
longer, always
wingless, and
usually in flowing
water

Common Pondskater

Gerris lacustris

The pondskater is a common and familiar
insect almost everywhere. They are frequently
to be seen out in the open, skating along on the
surface of a pond or river, casting shadows
where their feet rest on the surface film. This
species (generally the commonest of a group of
similar species) occurs in many kinds of water
bodies, from chalk streams to moorland pools.
It feeds on insects trapped in the surface film,
whose movements attract its attention. Some
individuals are winged, others are wingless,
with all grades in between. Hibernates during
winter; winged individuals usually go far away
from water. It is common and widespread
throughout most of Britain and Europe.

BUGS, HEMIPTERA

J	F	M	A	M	J
J	A	S	O	N	D

Common Froghopper
Philaenus spumarius

ID FACT FILE

SIZE:
Body length
5–6 mm

DESCRIPTION:
Nymphs are yellowish brown to greenish; adults variable, brownish-yellow with different patterns

FOOD:
Plant sap

Although the adult Common Froghoppers are not very well-known, their young nymph stages are much more familiar – the mass of frothy white liquid around them is the familiar 'cuckoo spit' of spring and early summer. The older insects emerge from the froth to live amongst soft herbaceous plants, on which they feed. They can occur in a wide variety of well-vegetated sheltered habitats, including gardens, and they are abundant throughout most of Britain and Europe.

| J | F | M | A | M | J |
| J | A | S | O | N | D |

Green Lacewing
Chrysopa carnea

ID FACT FILE

SIZE:
Body length
10–15 mm,
wingspan
35 mm

DESCRIPTION:
The body is pale
green and un-
marked. The
wings are mem-
branous and
clear green The
antennae are
long and fine

FOOD:
Insects

LOOKALIKES:
Chrysoperla perla
is very similar
and occurs in
similar habitats

One of several rather similar species known collectively as green lacewings. They are a distinctive and attractive group of delicate slow-moving insects, which often come into houses in autumn, attracted by the light. Their larvae are shuttle-shaped and bristly, and are voracious predators. They are especially fond of aphids, so they should be welcome in every garden. Eggs have a slender stalk and are laid under leaves. Winter is spent as a full-grown larva in a silk cocoon. This particular species mainly occurs in woodland or well-treed gardens and parks, and is widespread throughout Europe except the far north. In Britain, it is mainly southern in distribution. *Chrysoperla perla* is a similar species, with a pale green body marked with black and a black head; wings are slightly bluish-green, with some black cross-veins.

SCORPION FLIES, MECOPTERA

Scorpion Fly
Panorpa germanica

ID FACT FILE

SIZE:
Body length
17–20 mm

DESCRIPTION:
A distinctive
group of insects,
with a long ro-
bust downward-
pointed 'beak',
and the male's
upturned tail. The
wings are spot-
ted with black,
and the body is
yellowish

FOOD:
Dead and dying
insects and inver-
tebrates

J	F	M	A	M	J
J	A	S	O	N	D

Scorpion flies are so-called because the males have an upturned and curled over sting-like tail, resembling a scorpion, though they are quite harmless. Another characteristic feature is the beak-like projection used for feeding. They live as scavengers, or weak predators, including the skilful removal of prey from spiders' webs without themselves getting caught; they also look for food in ripe fruit. Their preferred habitat is woodland, hedgerows and other places with trees and shade. *P. germanica* is uncommon in Britain, but similar species *P. comunis* is more common and has heavy spotting on the wings. There are about 30 species of scorpion fly in Europe.

TRUE FLIES, DIPTERA

J	F	M	A	M	J
J	A	S	O	N	D

St Mark's Fly

Bibio marci

ID FACT FILE

SIZE:
Body length
about 10 mm

DESCRIPTION:
The body and
head are black
and furry. The
legs are black
and robust, and
the wings are
completely clear

FOOD:
Feeds mainly in
the larval stage,
though nectar is
taken by the
adults

LOOKALIKES:
B. hortulanus

The St. Mark's Fly is the only insect from this family that is at all familiar to the general public. It is known as the St. Mark's Fly because it appears at or about St. Mark's day (25th April), at the end of April. The flies settle on vegetation and flowers, and fly off slowly, trailing their legs, if disturbed. Their favoured habitats include woodland margins, grassy roadsides, hedgebanks, especially where flowering hawthorn trees are present. It is common throughout Britain and most of central and north Europe. *B. hortulanus* is similar in shape, but the females are orange-red, and less hairy; widespread, not uncommon in gardens in spring.

TRUE FLIES, DIPTERA

| J | F | M | A | M | J |
| J | A | S | O | N | D |

A Mosquito
Culex pipiens

ID FACT FILE

SIZE:
Body length
5–7 mm

DESCRIPTION:
A smallish mos-
quito, with grey-
brown body, and
white bands on
each abdominal
segment

FOOD:
Mammalian
blood

Mosquitoes are all-too-familiar as small whining flies which attack and suck blood, particularly at night or in dull weather. In fact, it is only the female which does this, requiring blood before the eggs can be laid. The males have feathery antennae, and hairy palps between the antennae. This is one of the commonest mosquitoes, though one of the least likely to bite, and it occurs everywhere; although the larvae live in water, almost any water will do, including a puddle. It occurs throughout Europe. There are several similar species; *Culiseta annulata* is even more boldly banded, but larger.

TRUE FLIES, DIPTERA

J	F	M	A	M	J
J	A	S	O	N	D

Chrysops relictus

ID FACT FILE

SIZE:
Body length
10 mm

DESCRIPTION:
The abdomen is
broad, yellowish
with black mark-
ings; eyes are
bright iridescent
green

FOOD:
Females drink
the blood of
mammals

These distinctive little flies are often admired
at first for their attractive shape and colours –
until they pierce the skin, when a painful
wound ensues. They are closely related to the
other horseflies, though are noticeably
different in holding their wings at rest in a
triangle. The females fly very quietly and
settle inconspicuously. The larvae live in wet
soil and mud, and the adults are most likely to
be found in damp undisturbed sites such as
woods and wet heaths. They are widespread
through north and central Europe, but not as
common in Britain.

J	F	M	A	M	J
J	A	S	O	N	D

A Hoverfly
Episyrphus balteatus

ID FACT FILE

SIZE:
Body length
10–12 mm

DESCRIPTION:
The thorax is
dark brown, and
the narrow abdo-
men is marked
with double black
stripes on yellow,
with the upper
stripe bolder
than the lower

FOOD:
Nectar as adult,
aphids as larvae

LOOKALIKES:
Nothing looks
quite like it

This is probably the commonest and best-
known of the hoverflies, partly because it is a
frequent visitor to parks and gardens as well as
more natural habitats, but also because it can
occur in enormous numbers. The adults are
very mobile and migratory, and can appear in
vast swarms in some years, especially if the
winds are from the south. The larvae are
voracious aphid-feeders, and the adults may be
found almost anywhere that there are flowers
for nectar. It occurs throughout Europe.
Syrphus ribesii is another hoverfly which
commonly visits parks and gardens as well as
natural habitats. Behaviour is as *Episyrphus
baleatus*, feeding on nectar both as adults and
larvae.

| J | F | M | A | M | J |
| J | A | S | O | N | D |

ID FACT FILE

SIZE:
Body length
10–12 mm

DESCRIPTION:
The thorax is
dark brown, and
the narrow
abdomen is
marked with dou-
ble black stripes
on yellow, with
the upper stripe
bolder than the
lower

FOOD:
Ants and other
prey

LOOKALIKES:
Wood Tiger
Beetle *C. sylvati-
ca* is larger and
darker than *hybri-
da* with a purple
tinge; yellow
markings are
less heavy

Tiger Beetle
Cicindela hybrida

Tiger beetles (*Cicindela* spp) are active
ground-dwelling predators, normally visible
during the daytime. If disturbed they fly for a
short distance before settling on the ground
again. The larvae live in vertical burrows in the
ground, holding themselves at the top in
readiness for any prey; once something is
caught, they take it to the base of the chamber
to eat it. *C. hybrida,* pictured here, has a local
scattered distribution on heathlands and dunes
in Britain. The more common and more
distinctive Green Tiger Beetle (*C. campestris*)
is a bright metallic green with small number of
yellow spots on its back (elytra). They are
found in open sunny habitats.

BEETLES, COLEOPTERA

Seven-spot Ladybird
Coccinella 7-punctata

ID FACT FILE

SIZE:
Body length
7–9 mm

DESCRIPTION:
Distinctive red
oval body, with
seven black dots
on the elytra.
The pronotum is
black with two
white dots.
Larvae are grey-
ish-blue with yel-
low dots

FOOD:
Aphids

LOOKALIKES:
No other
common species
normally has
seven spots

This is the most familiar of ladybirds, and the
one that is generally thought of as the typical
ladybird. When handled, they exude a strong-
smelling, acrid, yellowish fluid as a defence
against predators, reinforced by their bright
warning colours. Both adults and larvae are
active aphid hunters, scouring a wide range of
plants for food, and are very welcome in
gardens and amongst crops. The adults
hibernate, often in clusters in leaf litter or
sheltered hollows close to the ground. It is
common and widespread throughout most of
Britain and Europe, and migrates widely in
high population years.

BEETLES, COLEOPTERA

A Soldier Beetle
Cantharis rustica

ID FACT FILE

SIZE:
Body length
12–16 mm

DESCRIPTION:
They have red
and black heads,
a red thorax with
a black spot, and
dark bluish-black
soft elytra; the
legs are mainly
red

FOOD:
Small insects
and other inverte-
brates

LOOKALIKES:
C. fusca is simi-
lar with black
legs and a black
spot near the
front of the
thorax. Similar
habitats

The soldier beetles are a small group of
attractive beetles, so-called because of their
smart colours, reminiscent of an old-fashioned
soldier's uniform. They are active predators,
hunting on flowers and leaves in sunny
weather, looking for small insects. They can
also fly well at times, though tend not to. They
occur commonly in rough flowery places such
as woodland margins and glades, hedgerows
and scrub, and are widespread throughout
most of Britain and Europe, except the far
north.

BEETLES, COLEOPTERA

Whirligig Beetles
Gyrinus natator

ID FACT FILE

Size:
Body length
7–10 mm

Description:
Blackish oval,
with a finely stri-
ated elytra. Two
rear pairs of legs
are very short,
like little oars

Food:
Carrion, surface-
dwelling insects

Lookalikes:
Hairy Whirligig
*Orectochilus
villosus*

| J | F | M | A | M | J |
| J | A | S | O | N | D |

The whirligig beetles are very familiar as the
small black beetles that whirl around,
apparently aimlessly, on the surface of water in
groups. Their black, flattened legs are designed
for this purpose. They have eyes with two
parts, one for seeing above water and one for
seeing below water level. They feed on carrion
and insects at the water surface, including
mosquito larvae. The eggs are laid on
submerged plants, and the larvae are aquatic.
Whirligig beetles are most frequent in well-
vegetated still and slow-moving water bodies,
throughout Britain and Europe. Hairy
Whirligig *Orectochilus villosus* is slightly
larger, hairier and nocturnal.

ANTS, BEES, WASPS AND SAWFLIES, HYMENOPTERA

J	F	M	A	M	J
J	A	S	O	N	D

Black Garden Ant
Lasius niger

ID FACT FILE

SIZE:
Body length
4–9 mm

DESCRIPTION:
Black ant with a
single segment
at the waist

FOOD:
Omnivorous, with
a preference for
aphids

LOOKALIKES:
Yellow Meadow
Ant *L. flavius*, is
yellowish-brown-
but otherwise
similar

An extremely common ant often found on
pavements and garden paths. Sometimes
found nesting in or under house walls, and
under paving slabs. Although they are
ominorous they will 'milk' aphids for
honeydew. They may also make collars of soil
around plants to protect the aphids. Both this
species and the similar Yellow Meadow Ant
(*L. flavius*), produce huge mating swarms in
warm weather. The queens have wings and are
larger than the workers, but the wings break
off after mating. Mating flights occur in July or
August. Yellow Meadow Ants occur in rough
grassland, producing the familar ant-hills.

J	F	M	A	M	J
J	A	S	O	N	D

Wood Ant
Formica rufa

ID FACT FILE

SIZE:
Body length
10–12 mm

DESCRIPTION:
All stages are
largely black,
though the work-
ers (the most
commonly seen
stage) have a
red thorax and
red legs. The
nests are huge
conical piles of
plant material
such as pine
needles

FOOD:
Omnivorous, with
a preference for
invertebrate food

Wood Ants are amongst the most familiar and
distinctive of ants, partly because they are large
and occur in considerable abundance, but also
because they build huge nests. They live in
large colonies, and build nests of vegetation
which may reach 2m in height. There are
several queens in each nest. From these, the
workers fan out in vast numbers, often on
clearly defined paths, to seek food and nesting
material. Wood ants do not sting, but can spray
formic acid from their rear ends if alarmed.
They are protected in some countries because
of their value in reducing forest pests. They are
widespread and common in north-central
Europe and mountain areas further south,
usually in coniferous woods. There are
several rather similar wood ants, including
F. polyctena and *F. pratensis*.

J	F	M	A	M	J
J	A	S	O	N	D

German Wasp
Vespula germanica

ID FACT FILE

SIZE:
Body length:
12–20 mm, in-
cluding the larger
queens

DESCRIPTION:
Familiar black
and yellow band-
ed insects, with
a blackish tho-
rax. The mark-
ings on the tho-
rax and abdomen
are important in
distinguishing dif-
ferent species

FOOD:
Initially insects,
gradually chang-
ing over to fruit
and nectar

LOOKALIKES:
Several similar
species;
Common Wasp
V. vulgaris differs
in small details,
e.g. in having an
anchor of black
on its face, not
three black dots.
Similar distribu-
tion

This is one of the familiar wasps that come to seek out sweet substances in late summer, often coming into conflict with people in the process. In fact, there are several similar species known collectively as 'wasps' by the general public (see LOOKALIKES). They are social insects, living in colonies of hundreds or thousands of individuals, and it is the workers from the colonies that are most often seen. The nest itself is a delicate structure made from paper (chewed up wood), roughly spherical in shape. Young wasps are reared on insects collected by workers; adults feed on nectar and sweet material. Colonies break up in autumn, and only females survive the winter. They are common and widespread in a variety of habitats, including gardens, throughout Britain and Europe.

J	F	M	A	M	J
J	A	S	O	N	D

Bombus pascuorum

ID FACT FILE

SIZE:
9–18mm

DESCRIPTION:
Reddish-brown thorax; abdomen a mixture of brown, black or grey hairs. No white tip to abdomen

FOOD:
Nectar and pollen

LOOKALIKES:
On *B. lucorum* the collar of thorax and second abdominal segment is yellow, and the tip of the abdomen is white; *B hortorum* has a yellow band where thorax and abdomen meet, also has white tip to abdomen; *B. terrestris* has same markings as *B. lucorum*, but bands are darker orange or gold

Bumblebees are very familiar as a group, though few people realise how many different species there are in Europe, and they can be quite difficult to tell apart. *Bombus pascuorum* pictured here is a common garden species throughout Britain and Europe. It is found in most habitats apart from more exposed areas. It is characterised by a reddish brown thorax, and is the only garden species with this colour. It visits a number of different flowers, both wild and cultivated, but particularly favours lavender. Nests are usually built at ground level in long grass; this species may also build in old birds' nests above ground. Colonies persist well into the autumn, after other species have disappeared.

HESPERIIDAE (SKIPPERS)

J	F	M	A	M	J
J	A	S	O	N	D

Large Skipper
Ochlodes venatus

ID FACT FILE

WINGSPAN:
2.5–3.2 cm

DESCRIPTION:
Uppersides very similar to Silver-spotted Skipper, but female spotting much darker. Undersides also similar but lacking the silvery spots

HIBERNATING STAGE:
Caterpillar

FLIGHT PERIOD:
Summer

CATERPILLAR FOOD PLANTS:
Various grasses

LOOKALIKES:
Silver-spotted Skipper

This butterfly is common throughout most of Europe, but is absent from the northern parts of the continent, including Scotland. It is usually found on grassy hillsides, and also around forest edges and roadsides, up to about 2000 m. The green caterpillar is striped with darker green down the back; it is yellow at the sides and has a large dark brown head. It pupates within a cocoon amongst grass blades. Up to three broods in the south.

male

female

PIERIDAE (WHITES AND YELLOWS)

J	F	M	A	M	J
J	A	S	O	N	D

Orange Tip
Anthocharis cardamines

ID FACT FILE

WINGSPAN:
3.8–4.8cm

DESCRIPTION:
White with black
tips to the
forewings, male
with large orange
patch. Underside
mottled with
black and yellow

**HIBERNATING
STAGE:**
Pupa

FLIGHT PERIOD:
Mid- to late
spring

**CATERPILLAR FOOD
PLANTS:**
Lady's smock
(*Cardamine*),
hedge mustard
(*Sisymbrium*),
and related
plants

This butterfly is common throughout Europe
except S Spain and the far north. It prefers
open woodland, flowery meadows and sometimes gardens, up to 2000 m. The green caterpillar has a white line along each side
and is darker underneath. It feeds on the
flowers and seed-pods of its food plants, and
is also cannibalistic, eating any other eggs and
caterpillars it finds on the same plant. The pupa
is very slender and looks a little like a seed-pod.

male

female

PIERIDAE (WHITES AND YELLOWS)

J	F	M	A	M	J
J	A	S	O	N	D

Brimstone
Gonepteryx rhamni

ID FACT FILE

WINGSPAN:
5.2–6 cm

DESCRIPTION:
Male bright
yellow, female
greenish-white,
with an orange
spot in the mid-
dle of each wing.
Underside similar
but greener

**HIBERNATING
STAGE:**
Adult

FLIGHT PERIOD:
Early spring to
late summer,
new adults
hatching in early
summer

**CATERPILLAR FOOD
PLANTS:**
Buckthorn
(*Rhamnus*)

LOOKALIKES:
Cleopatra

This common butterfly is found throughout
Europe except Scotland and N Scandinavia.
It prefers open woodland and forest edges,
sometimes gardens, up to around 1800 m, but,
like many butterflies in this family, it keeps on
the move, never staying in one place for long.
The caterpillar is green with a pale line along
each side, and lies along the middle of a leaf
when not feeding, which makes it very hard to
find. The pupa is very leaf-like.

female

▲ male underside

PIERIDAE (WHITES AND YELLOWS)

J	F	M	A	M	J
J	A	S	O	N	D

Large White
Pieris brassicae

ID FACT FILE

WINGSPAN:
5.6–6.6 cm

DESCRIPTION:
White with forewing tips black. Female has 2 black spots on forewings. Underside similar but hindwings pale greyish-green

HIBERNATING STAGE:
Pupa

FLIGHT PERIOD:
Mid-spring to late summer; 2 or 3 broods

CATERPILLAR FOOD PLANTS:
Nasturtium (*Tropaeolum*), cabbage (*Brassica*), and related plants

This is probably the most familiar butterfly in Europe, found everywhere except the far north, although in summer it migrates even there. It is found wherever there are flowers and suitable food plants, which has made it something of a pest in gardens and agricultural areas. The caterpillars are green with black spots and a yellow line along the back and sides; they live in a group when young and often wander some distance before pupating.

left male
right female

LYCAENIDAE (BLUES, COPPERS AND HAIRSTREAKS)

Common Blue

Polyommatus icarus

J	F	M	A	M	J
J	A	S	O	N	D

ID FACT FILE

WINGSPAN:
2.5–3 cm

DESCRIPTION:
Male light violet-blue, very finely edged with black; female dark brown, often with some blue close to the body and with orange crescents near the outer edges of all 4 wings. In both sexes, underside pale brownish-grey with blue near the body, marked with orange crescents near edges of hindwings, and black spots

HIBERNATING STAGE:
Caterpillar

FLIGHT PERIOD:
Mid-spring to late summer; 2 or 3 broods

CATERPILLAR FOOD PLANTS:
Bird's-foot-trefoil (*Lotus*), clover (*Trifolium*) and many other related plants

This is one of the commonest butterflies, as its name may suggest, and is found throughout Europe, preferring meadows and flowery hillsides up to around 2500 m, although it is sometimes seen in gardens. The caterpillar is green with a darker line down the back and yellowish stripes along the sides. Like many blues, they show a preference for flowers when feeding, and are normally attended by ants. Pupation takes place at ground level.

male

female

J	F	M	A	M	J
J	A	S	O	N	D

Small Copper
Lycaena phlaeas

ID FACT FILE

WINGSPAN:
2.2–2.7 cm

DESCRIPTION:
Forewings
shining coppery-
orange with dark
brown edges and
black spots;
hindwings dark
brown with cop-
pery band close
to outer edges.
Underside pale
grey-brown with
black spots.
Sexes alike

**HIBERNATING
STAGE:**
Caterpillar

FLIGHT PERIOD:
Early spring to
early autumn

**CATERPILLAR FOOD
PLANTS:**
Dock and sorrel
(*Rumex*), and
knotgrass
(*Polygonum*)

caterpillar

This very common butterfly is found throughout Europe, and prefers sunny meadows and
forest clearings up to around 2000 m. Males
are extremely territorial, chasing any intruders
away, even birds! There are normally two,
sometimes three broods each year, third
generation adults are often much smaller. The
caterpillar is green, sometimes with a dark pink
stripe down the back and sides, and is attended
by ants. It usually pupates under a leaf of the
food plant.

J	F	M	A	M	J
J	A	S	O	N	D

Small Tortoiseshell
Aglais urticae

ID FACT FILE

WINGSPAN:
4.5–5 cm

DESCRIPTION:
Orange-red
ground colour
with black mark-
ings, edged
black with blue
crescents.
Underside dark,
almost black.
Sexes alike

**HIBERNATING
STAGE:**
Adult

FLIGHT PERIOD:
Spring, and early
summer to
autumn

**CATERPILLAR FOOD
PLANTS:**
Nettle (*Urtica*)

LOOKALIKES:
Large Tortoise-
shell

Found throughout Europe, this is probably one
of the most popular butterflies seen in gardens,
where they favour *Buddleia* and *Sedum* flowers
at the height of summer. They are also often
found indoors in the winter when a warm spell
might rouse them from hibernation. Males are
territorial. There are two broods each year
except in the far north. Caterpillars live in communal webs when young, for protection against
predators. Shortly before pupating, they will
separate and feed alone.

caterpillars

| J | F | M | A | M | J |
| J | A | S | O | N | D |

Map Butterfly
Araschnia levana

ID FACT FILE

WINGSPAN:
3.2–4 cm

DESCRIPTION:
First generation,
light orange-
brown with dark
brown and white
spots; second,
dark brown with
white bands.
Underside red-
dish-brown with
white bands and
yellowish edges.
Sexes alike

**HIBERNATING
STAGE:**
Pupa

FLIGHT PERIOD:
Late spring and
late summer

**CATERPILLAR FOOD
PLANTS:**
Nettle (*Urtica*)

This little butterfly is common throughout the
lowlands of central and E Europe, preferring
open woodland and forest edges. There are
two broods each year, and the adults from the
spring generation are remarkably different
from those of the summer. Eggs are laid in
strings under leaves, and the young caterpillars
live in groups, separating when nearly fully
grown. They are mainly black with rows of
black or brown branching spines.

underside of
2nd brood

left male 1st brood
right male 2nd brood

NYMPHALIDAE (BRUSH-FOOTED BUTTERFLIES)

J	F	M	A	M	J
J	A	S	O	N	D

Queen of Spain Fritillary
Argynnis lathonia

ID FACT FILE

WINGSPAN:
3.8–5 cm

DESCRIPTION:
Bright orange-brown with black spots. Underside slightly paler, with black spots on forewings, hindwings with very large silver spots. Sexes alike

HIBERNATING STAGE:
Egg, caterpillar, pupa or adult, depending on locality

FLIGHT PERIOD:
Early spring to early autumn

CATERPILLAR FOOD PLANTS:
Violet (*Viola*)

A common butterfly in S Europe, but also a well-known migrant that spreads to most of the rest of the continent every summer, reaching as far north as S Scandinavia. It is, however, a very rare visitor to the British Isles. It prefers meadows and heaths up to around 2500 m, and in the south there may be two or three broods each year. The fully grown caterpillar pupates under a leaf on the food plant.

caterpillar

Painted Lady
Cynthia cardui

ID FACT FILE

WINGSPAN:
5.4–5.8 cm

DESCRIPTION:
Pale pinkish-brown with black spots, also white spots on forewings. Underside marbled pale and mid-brown with a row of blue eye-like spots on hind-wings. Sexes alike

HIBERNATING STAGE:
Adult, but only in extreme south

FLIGHT PERIOD:
Mid-spring to late autumn

CATERPILLAR FOOD PLANTS:
Thistle (*Carduus*) and nettle (*Urtica*)

This is not a true resident in Europe, as it migrates from Africa every year and is unable to survive the winter anywhere else except, perhaps, in the extreme south. However, it is a common sight everywhere during the summer, even in parks and gardens, up to around 2500 m. Migrants produce one, sometimes two broods each year. The young caterpillar usually makes a shelter out of one or two leaves, but will often stop doing this when much larger. It pupates inside a similar shelter.

caterpillar

NYMPHALIDAE (BRUSH-FOOTED BUTTERFLIES)

J	F	M	A	M	J
J	A	S	O	N	D

Peacock
Inachis io

ID FACT FILE

WINGSPAN:
5.4–5.8 cm

DESCRIPTION:
Dark reddish-brown with black and yellow markings and a large lilac-blue eye-like spot on each wing. Underside mostly black. Sexes alike

HIBERNATING STAGE:
Adult

FLIGHT PERIOD:
Mid-summer to early autumn and early to mid-spring

CATERPILLAR FOOD PLANTS:
Nettle (*Urtica*)

This very common butterfly is found throughout Europe except N Scotland and N Scandinavia. It can be seen anywhere there is a good supply of suitable flowers, especially parks and gardens, up to around 2000 m. The caterpillars live in groups when young, separating when nearly fully grown, and will often wander some distance from the food plant before pupating. Adults will often enter houses in the search for somewhere to pass the winter.

caterpillar

NYMPHALIDAE (BRUSH-FOOTED BUTTERFLIES)

Red Admiral
Vanessa atalanta

ID FACT FILE

WINGSPAN:
5.6–6.2 cm

DESCRIPTION:
Black with red bands and white spots. Underside mostly dark brown mottled black and blue. Sexes alike

HIBERNATING STAGE:
Adult, rarely surviving winter in the north

FLIGHT PERIOD:
Late spring to early autumn and early to mid-spring

CATERPILLAR FOOD PLANTS:
Nettle (*Urtica*)

This common butterfly is resident only in S Europe, but every year it migrates north-wards, reaching every part of the continent. It can be found wherever there are flowers, and is a common sight in gardens. In the autumn it is strongly attracted to rotten fruit. The cater-pillar lives inside a folded leaf all through its development, and when fully grown often wanders some distance from the food plant before pupating.

caterpillar

J	F	M	A	M	J
J	A	S	O	N	D

Ringlet
Aphantopus hyperantus

ID FACT FILE

Sub-family:
Satyrinae (Satyrs and wood nymphs)

Wingspan:
4–4.8 cm

Description:
Very dark brown with 1 or 2 black eye-like markings on each wing. Underside dark brown with several yellow-ringed black eye-spots. Sexes alike

Hibernating stage:
Caterpillar

Flight period:
Early summer

Caterpillar food plants:
Various grasses

This butterfly is common throughout most of Europe, but is absent from N Scandinavia and N Scotland, also most of Spain and Italy. It prefers open woodland, hedgerows and damp meadows up to around 1500 m, and is very fond of bramble (*Rubus*) flowers. The female scatters her eggs while flying over grassy areas, and the resulting caterpillars feed mainly at night. When fully grown they pupate on the ground in a flimsy cocoon.

caterpillar

NYMPHALIDAE (BRUSH-FOOTED BUTTERFLIES)

| J | F | M | A | M | J |
| J | A | S | O | N | D |

Small Heath
Coenonympha pamphilus

ID FACT FILE

SUB-FAMILY:
Satyrinae (Satyrs and wood nymphs)

WINGSPAN:
2.6–3.2 cm

DESCRIPTION:
Light orange-brown with brown edges and a black spot on each forewing. Underside pale brownish-grey with a short white band and an eye-spot on each forewing. Female larger

HIBERNATING STAGE:
Caterpillar

FLIGHT PERIOD:
Mid-spring to early autumn, depending on locality

CATERPILLAR FOOD PLANTS:
Various grasses

LOOKALIKES:
Large Heath and Chestnut Heath

This very common butterfly is found throughout Europe except the far north, and prefers meadows, heaths and any other kind of rough grassy areas, up to around 2000 m. There may be up to three broods each year, depending on locality, but some caterpillars from each of the first broods will hibernate along with all those of the last brood. The caterpillar usually feeds during the day, and later attaches itself to a grass stem to pupate.

caterpillar

J	F	M	A	M	J
J	A	S	O	N	D

Grayling
Hipparchia semele

ID FACT FILE

Sub-Family:
Satyrinae (Satyrs and wood nymphs)

Wingspan:
4.2–6 cm

Description:
Mid-brown with a paler band and 2 eye-spots on the forewings, 1 on the hindwings. Underside similar but with a whitish band on the hindwings. Female darker, bands on upper-side yellow

Hibernating stage:
Caterpillar

Flight period:
Mid-summer

Caterpillar food plants:
Various grasses

This butterfly is found throughout Europe except S Italy and N Scandinavia. In the British Isles it is mainly restricted to coastal areas. It prefers heathland, sand hills and open woodland, up to around 2000 m, and is very hard to see when it settles on a patch of bare earth. The caterpillar is pale brown with darker stripes along the back and sides. It feeds mainly at night, and pupates in a flimsy cocoon on the ground.

male

NYMPHALIDAE (BRUSH-FOOTED BUTTERFLIES)

Meadow Brown
Maniola jurtina

ID FACT FILE

SUB-FAMILY:
Satyrinae (Satyrs and wood nymphs)

WINGSPAN:
4.2–5.4 cm

DESCRIPTION:
Dark brown with 1 eye-spot on each forewing, female marked with orange-brown. Underside light brown with a paler band

HIBERNATING STAGE:
Caterpillar

FLIGHT PERIOD:
Late spring to late summer

CATERPILLAR FOOD PLANTS:
Various grasses

LOOKALIKES:
Dusky Meadow Brown

This variable butterfly is very common throughout Europe, but is absent from N Scandinavia and Finland. It is found in all grassy places, including open woodland and forest edges, up to around 2000 m, and is often seen flying even on dull days. The caterpillar feeds mainly at night and is green with a darker line down the back and a thin, pale line along the sides. The green pupa is attached to a grass stem.

left male
right female

NYMPHALIDAE (BRUSH-FOOTED BUTTERFLIES)

| J | F | M | A | M | J |
| J | A | S | O | N | D |

Wall Brown
Lasiommata megera

ID FACT FILE

Sub-Family:
Satyrinae (Satyrs and wood nymphs)

Wingspan:
3.6–5 cm

Description:
Dark brown with large orange-brown spots, one eye-spot on the forewings, and several on the hindwings. Underside pale brown with smaller eye-spots. Female paler

Hibernating stage:
Caterpillar

Flight period:
Early spring to late summer; 2 or 3 broods

Caterpillar food plants:
Various grasses

This is a common butterfly in most of Europe, but it is absent from Finland and N Scandinavia. It prefers heathland and other grasslands, open woodland and gardens, up to around 2000 m, and likes to bask in the sun on stones or open ground. The caterpillar is green with whitish lines along the back and sides, and feeds mainly at night. The pupa is usually green but may be marked with black, and is attached to a grass stem.

male

female

NYMPHALIDAE (BRUSH-FOOTED BUTTERFLIES)

Speckled Wood
Pararge aegeria

J	F	M	A	M	J
J	A	S	O	N	D

ID FACT FILE

Sub-Family:
Satyrinae (Satyrs and wood nymphs)

Wingspan:
3.8–4.4 cm

Description:
Dark brown with cream spots in the north, orange-brown spots in the south, and several eye-spots. Underside marbled light and dark brown. Sexes alike

Hibernating stage:
Both caterpillar and pupa

Flight period:
Early spring to early autumn, depending on locality

Caterpillar food plants:
Various grasses

A common butterfly throughout most of Europe, except the far north, with two distinct colour forms. It prefers shady woodland up to around 1200 m. Males are very territorial, and will fiercely defend a patch of sunlight. There may be two or three broods each year, depending on locality. The caterpillar is green with a darker stripe down the back and paler lines along the sides. The green pupa is attached to a grass stem.

northern race

left southern race

J	F	M	A	M	J
J	A	S	O	N	D

Gatekeeper
Pyronia tithonus

ID FACT FILE

SUB-FAMILY:
Satyrinae (Satyrs and wood nymphs)

WINGSPAN:
3.4–3.8 cm

DESCRIPTION:
Mid-brown with large orange-brown patches and 1 eye-spot on each forewing. Underside lighter, hindwing with several small eye-spots and marbled with pale brown. Female paler

HIBERNATING STAGE:
Caterpillar

FLIGHT PERIOD:
Mid-summer

CATERPILLAR FOOD PLANTS:
Various grasses

LOOKALIKES:
Southern Gatekeeper

Also known as the Hedge Brown, this very common butterfly is found throughout central and S Europe, also S England, Ireland and Wales. It prefers open woodland, hedgerows and grassland below 1000 m, and is very fond of bramble (*Rubus*) flowers. The caterpillar is pale brown with a darker stripe down the back and a pale line along the sides, and feeds mainly at night. The pale brown pupa is attached to a grass stem.

male

female

J	F	M	A	M	J
J	A	S	O	N	D

Magpie
Abraxas grossulariata

ID FACT FILE

Wingspan:
3.5–4 cm

Description:
Yellow body with yellow line across each forewing, otherwise wings clear white with variable black spots. Female larger

Hibernating st8age:

Caterpillar

Flight period:
Late spring to early summer

Caterpillar food plants:
Sloe (*Prunus*), hawthorn (*Crataegus*), gooseberry (*Ribes*) and other currant bushes

The Magpie is a common moth throughout Europe, sometimes very common indeed, and is quite an eye-catching sight, so brightly is it coloured. This bold pattern means, of course, that the moth is poisonous (i.e. tastes nasty), and therefore to be avoided by predators looking for a light snack. The moth, which can often be seen flying during the day, is very variable, and so is the caterpillar, which has the same colouring as the adult.

caterpillar

LASIOCAMPIDAE (EGGARS)

J	F	M	A	M	J
J	A	S	O	N	D

Lappet
Gastropacha quercifolia

ID FACT FILE

WINGSPAN:
5–9 cm

DESCRIPTION:
Ground colour usually rich reddish-brown but this may vary. Edges of the wings scalloped. At rest the moth looks like a bunch of dead beech leaves. Sexes alike, but female larger

HIBERNATING STAGE:
Caterpillar

FLIGHT PERIOD:
Summer

CATERPILLAR FOOD PLANTS:
A number of trees and shrubs including fruit trees, but particularly sloe (*Prunus*) and hawthorn (*Crataegus*)

A beautiful, well-camouflaged moth, becoming rarer owing to pesticides, although it can still be found throughout most of Europe except Ireland, Scotland and N Scandinavia. The name comes from the fleshy 'lappets' down the sides of the caterpillar which help to conceal it by reducing shadow. The caterpillar spends the winter on a twig while still small, then when fully grown it makes a cocoon attached to a branch in which to pupate.

caterpillar

| J | F | M | A | M | J |
| J | A | S | O | N | D |

Elephant Hawkmoth
Deilephila elpenor

ID FACT FILE

WINGSPAN:
5.8–7 cm

DESCRIPTION:
Deep pink with
bands of olive-
green; hindwings
marked with
black near body.
Legs and
antennae white.
Female larger

**HIBERNATING
STAGE:**
Pupa

FLIGHT PERIOD:
Early summer

**CATERPILLAR FOOD
PLANTS:**
Willowherb
(*Epilobium*);
fuchsia in
gardens

LOOKALIKES:
Small Elephant
Hawkmoth

This moth is common throughout Europe,
preferring open woodland, waste ground, parks
and gardens in lowland areas, wherever the
food plant grows. The odd name describes the
caterpillar, which has a tapering front end that
looks slightly like an elephant's trunk. This can
be withdrawn rapidly to puff up the alarming
false eye-spots. Caterpillars are often seen in
August moving quickly across open ground as
they look for somewhere to pupate.

caterpillar

SPHINGIDAE (HAWKMOTHS)

J	F	M	A	M	J
J	A	S	O	N	D

Poplar Hawkmoth
Laothoe populi

ID FACT FILE

WINGSPAN:
7–9 cm

DESCRIPTION:
Marbled shades of light and dark grey. Hindwings have a reddish-brown patch near the body. Female larger

HIBERNATING STAGE:
Pupa

FLIGHT PERIOD:
Late spring, also late summer in the south

CATERPILLAR FOOD PLANTS:
Poplar (*Populus*), sallow and willow (*Salix*)

This moth is very common throughout Europe, although perhaps less so in the south. It prefers open woodland, parks and gardens, mainly in damp lowland areas. The adults do not feed, but are easily attracted to lights. There are usually two broods each year in the south, but only one further north. The caterpillar is very well camouflaged, and when fully grown it pupates underground in a small chamber.

male

caterpillar

SPHINGIDAE (HAWKMOTHS)

Hummingbird Hawkmoth

Macroglossum stellatarum

ID FACT FILE

WINGSPAN:
4–5 cm

DESCRIPTION:
Forewings grey with darker wavy lines; hindwings light orange-brown. Tail of body black and white. Sexes alike

HIBERNATING STAGE:
Adult, sometimes pupa. Can only survive in the south

FLIGHT PERIOD:
All year

CATERPILLAR FOOD PLANTS:
Bedstraw (*Galium*) and valerian (*Centranthus*)

This moth is very common in S Europe, but migrates northwards every year, reaching as far as N Scandinavia and the British Isles. It can be found almost anywhere, but especially in parks and gardens. It flies during the day, and if it finds a well-stocked garden, it will stay for several days, patrolling regularly every four hours. There are two broods each year in S Europe, but adults may be seen all year. The fully grown caterpillar pupates in a flimsy cocoon on the ground amongst leaf-litter.

caterpillar

NOCTUIDAE (OWLET MOTHS)

Silver Y
Autographa gamma

ID FACT FILE

WINGSPAN:
3.5–4.3 cm

DESCRIPTION:
Forewings light brownish-grey mottled with darker shades and marked with a silvery-white 'Y'. Hindwings pale brown. 2 prominent tufts of hair on the back. Sexes alike

HIBERNATING STAGE:
Caterpillar

FLIGHT PERIOD:
Late spring to autumn;
2 broods

CATERPILLAR FOOD PLANTS:
Many low-growing plants

This moth is mainly resident in S Europe, but migrates northwards every year; it can reach as far north as Lapland, but cannot survive the winter anywhere. It is found almost everywhere, having no strong preferences except a good supply of flowers. It can be seen flying during both day and night, but most particularly at dusk. The caterpillar varies in colour from light to dark green, and pupates in a cocoon amongst leaf-litter.

caterpillar

NOCTUIDAE (OWLET MOTHS)

J	F	M	A	M	J
J	A	S	O	N	D

Red Underwing
Catocala nupta

ID FACT FILE

WINGSPAN:
7–8 cm

DESCRIPTION:
Forewings light grey-brown with a darker mottled pattern; hind-wings banded black and scarlet with a white fringe

HIBERNATING STAGE:
Egg

FLIGHT PERIOD:
Late summer

CATERPILLAR FOOD PLANTS:
Poplar (*Populus*) and related trees

This moth is common throughout most of Europe, but is absent from N Scandinavia, Scotland and Ireland. It prefers open woodland, parks and gardens and is extremely hard to spot when resting on a tree trunk, but at night it is strongly attracted to rotten fruit. The caterpillar feeds only at night, and when fully grown pupates in a cocoon either amongst leaf-litter on the ground or under bark.

caterpillar

NOCTUIDAE (OWLET MOTHS)

Large Yellow Underwing
Noctua pronuba

This very common moth is found throughout Europe, and can be seen almost everywhere, especially in gardens and other cultivated land where it may sometimes become a pest. It is easily attracted to flowers and bright lights at night, but in the warmer south it may become dormant during the hottest weather. The caterpillar may be green or brown; it feeds at night, hiding underground during the day, and pupates below ground when fully grown.

ID FACT FILE

WINGSPAN:
5–6 cm

DESCRIPTION:
Forewings mid-brown with 1 darker and 1 paler spot near the middle, and another darker spot near the tip; hindwings bright yellow with a black band near the edges. Female larger and paler

HIBERNATING STAGE:
Caterpillar

FLIGHT PERIOD:
Early summer to early autumn

CATERPILLAR FOOD PLANTS:
Many low-growing plants

caterpillar

| J | F | M | A | M | J |
| J | A | S | O | N | D |

Herald
Scoliopteryx libatrix

ID FACT FILE

WINGSPAN:
4–4.5 cm

DESCRIPTION:
Pale reddish-brown; forewings deeply scalloped, with orange-brown patches and white lines. Sexes alike

HIBERNATING STAGE:
Adult

FLIGHT PERIOD:
Late summer to early autumn and spring

CATERPILLAR FOOD PLANTS:
Poplar (*Populus*) and willow (*Salix*)

This common moth can be found everywhere in Europe, preferring damp, open places, and also parks and gardens up to around 2000 m. It is usually seen feeding from flowers such as ivy (*Hedera*), but can sometimes be attracted to very ripe fruit. Occasionally it can be found trying to hibernate in a shed or other outhouse. The caterpillar is extremely well camouflaged, and when fully grown makes a white cocoon amongst the leaves in which to pupate.

caterpillar

ARCTIIDAE (TIGER AND FOOTMAN MOTHS)

J	F	M	A	M	J
J	A	S	O	N	D

Garden Tiger
Arctia caja

ID FACT FILE

WINGSPAN:
4.5–6.5 cm

DESCRIPTION:
Very variable. Forewings usually white with dark brown spots; hindwings red to yellow with blue-centred black spots. Female larger

HIBERNATING STAGE:
Caterpillar

FLIGHT PERIOD:
Early to mid-summer

CATERPILLAR FOOD PLANTS:
Nettle (*Urtica*), dandelion (*Taraxacum*) and many other low-growing plants

LOOKALIKES:
Cream-spot Tiger

A very common moth throughout Europe, and can be found almost anywhere that suitable food plants can be found. It is very variable in appearance, so much so, in fact, that it is rare to find two individuals with identical markings. The densely hairy caterpillars can be seen feeding up in the spring, particularly on sunny days, and are commonly known as 'woolly bears'. Pupation takes place within a white silk cocoon.

caterpillar

ARCTIIDAE (TIGER AND FOOTMAN MOTHS)

Ruby Tiger
Phragmatobia fuliginosa

ID FACT FILE

WINGSPAN:
3.4–4 cm

DESCRIPTION:
Dark reddish-brown, hindwings mainly black. Two small black spots on each wing. Female larger

HIBERNATING STAGE:
Caterpillar

FLIGHT PERIOD:
Late spring to early summer, also late summer to early autumn in the south

CATERPILLAR FOOD PLANTS:
Many low-growing plants such as dock (*Rumex*)

This fairly common moth is found almost everywhere in Europe, although in some areas it is only locally common. It can be seen in most habitats, particularly open woodland and meadows, anywhere that suitable food plants can be found. There are two broods each year in the south, but normally only one in the cooler north. In spring, after hibernation, the fully grown caterpillar pupates in a cocoon on the ground, usually amongst leaf-litter.

caterpillar

ARCTIIDAE (TIGER AND FOOTMAN MOTHS)

White Ermine
Spilosoma lubricepeda

ID FACT FILE

WINGSPAN:
3–4.2 cm

DESCRIPTION:
Variable pattern of small black spots on white ground. Easily recognised from any other white moth by the yellow abdomen. Sexes alike.

HIBERNATING STAGE:
Pupa

FLIGHT PERIOD:
Late spring to early summer

CATERPILLAR FOOD PLANTS:
Almost any low-growing plant

This very common moth is found almost everywhere in Europe, and in virtually every kind of habitat, although it does have a preference for open spaces. It is easily attracted to lights at night, and this is probably when it is most often seen. The caterpillar makes a cocoon close to the ground amongst leaves or leaf-litter. There is occasionally a partial second brood in the autumn, when the climate is favourable.

rups

caterpillar

ARCTIIDAE (TIGER AND FOOTMAN MOTHS)

J	F	M	A	M	J
J	A	S	O	N	D

Cinnabar
Tyria jacobaeae

ID FACT FILE

WINGSPAN:
3.2–4.2 cm

DESCRIPTION:
Forewings black
with a red line
close to the lead-
ing edge and
2 red spots;
hindwings red,
edged with black.
Sexes alike

**HIBERNATING
STAGE:**
Pupa

FLIGHT PERIOD:
Late spring to
early summer

**CATERPILLAR FOOD
PLANTS:**
Ragwort
(*Senecio*)

A moth of open places, such as heathland, and
downland, and found through most of Europe,
although it is not abundantly common. Mainly
a night-flying species, it is easily disturbed
during the day, when its weak flight and bright
colours make it easy to see. The bright colours
of both adult and caterpillar advertise the fact
that this species is distasteful to birds. The
caterpillar makes a flimsy cocoon on or under
the ground.

caterpillar

ZYGAENIDAE (BURNETS AND FORESTERS)

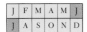

J	F	M	A	M	J
J	A	S	O	N	D

Six-spot Burnet
Zygaena filipendulae

ID FACT FILE

WINGSPAN:
3–3.8 cm

DESCRIPTION:
Black with a bluish sheen and 6 red spots on each forewing; hindwings red with black edges. Sexes alike

HIBERNATING STAGE:
Caterpillar

FLIGHT PERIOD:
Early summer

CATERPILLAR FOOD PLANTS:
Bird's-foot-trefoil (*Lotus*)

This is probably the commonest of the burnet moths, and is found throughout most of Europe except the far north and S Spain. It prefers flowery meadows and other grasslands, up to around 2000 m. The caterpillar is pale green with black spots and a black head, and pupates in a straw-coloured cocoon high up on a grass stem. Both adult and caterpillar can produce a poisonous liquid from special glands to warn off predators.

mating pair

six-spot Burnet
exuding toxins

J	F	M	A	M	J
J	A	S	O	N	D

Smooth Newt

Triturus vulgaris

ID FACT FILE

TOTAL LENGTH:
Up to 11cm,
but usually less

FIELD CHARACTERS:
Breeding male
heavily spotted.
Both sexes have
reddish centre to
belly

LARVAE:
Up to 4cm with
tapering tail;
external gills

SIMILAR SPECIES:
Other small
newts, which are
generally not as
heavily spotted
on the underside

Widespread and often the commonest newt over most of Europe, except the south-west, occurring in almost any habitat near to its breeding waters. Like other newts it feeds on invertebrates, also probably fish fry and amphibian eggs. It is a relatively small newt and in the breeding season the male develops a very large, wavy crest along the body and tail. Like most other newts, it leaves the water after breeding, and the skin becomes dryish and velvety, and might be confused with a lizard. It normally hides under logs and stones.

J	F	M	A	M	J
J	A	S	O	N	D

Common Toad

Bufo bufo

ID FACT FILE

TOTAL LENGTH:
Up to 15cm

TAIL LENGTH:
Absent

FIELD CHARACTERS:
Large, warty

SOUNDS:
A quiet
quark-quark-quark

BREEDING:
In water, laying
eggs in strings

SIMILAR SPECIES:
Other toads,
which are
generally more
colourful

A relatively large, robust toad with an extremely warty skin and large parotoid glands. In the south of its range the warts are sometimes spiny. Like all true toads and frogs the pupil is a horizontal slit in bright light, and its colour changes slightly to match its surroundings. It occurs in a very wide range of habitats, feeding on invertebrates and other animals up to the size of small mice. During the breeding season males gather in shallow water, and have a rather soft call. The eggs are laid in long strings, usually wrapped around water weeds at the bottom of a pond or lake. The young emerge from the water usually after rain, often in huge numbers.

J	F	M	A	M	J
J	A	S	O	N	D

Natterjack
Bufo calamita

ID FACT FILE

TOTAL LENGTH:
Up to 10cm, but
usually smaller

TAIL LENGTH:
Absent

FIELD CHARACTERS:
Yellow stripe
down back

SOUNDS:
Usually heard
in chorus

BREEDING:
In water, eggs
in strings

SIMILAR SPECIES:
Green Toad,
which has
brighter colouring

A small, relatively short-limbed toad that tends to run rather than hop. It is distinguished from other toads by the thin yellow stripe down the middle of the back. In Britain it is largely confined to sandy habitats. There is a small population in south west Ireland, probably as a result of human introduction. Often nocturnal, spending the day in burrows, which it excavates. During the breeding season males can be heard for 2km or more – each call is a rapidly repeated ratchet-like sound, but when several hundred call they merge into a continuous sound.

FROGS AND TOADS

J	F	M	A	M	J
J	A	S	O	N	D

Common Frog
Rana temporaria

ID FACT FILE

TOTAL LENGTH:
Up to 10cm

TAIL LENGTH:
Absent

FIELD CHARACTERS:
Brownish (but variable) with dark blotches

SOUNDS:
A *grook-grook-grook*, under water

BREEDING:
Lays eggs in large clumps that float on the surface when freshly laid

SIMILAR SPECIES:
Moor Frog; Agile Frog

The Common Frog is one of the most widespread and often most abundant amphibian in Europe; it is extremely tolerant of the cold and occurs in almost all habitats with suitable breeding waters, even as far north as the Arctic Circle. The colouring is very variable, and can be yellowish, brown or reddish. There is a characteristic dark stripe between the nostrils and the eyes. Compared with other frogs, it has relatively short hind limbs, but it still a powerful jumper. During the breeding season the male has black thumb pads for clasping the female; the call is rather quiet and produced under water.

FROGS AND TOADS

J	F	M	A	M	J
J	A	S	O	N	D

Edible Frog
Rana esculenta

ID FACT FILE

TOTAL LENGTH:
Up to 9cm

TAIL LENGTH:
Absent

FIELD CHARACTERS:
Green with black blothes; relatively long hindlegs mottled with black, yellow or orange

SOUND:
Highly vocal; rasping calls can carry a long distance

BREEDING:
Eggs laid in clumps that do not float

SIMILAR SPECIES:
Marsh Frog, Pool Frog; other 'green' frogs

For most of this century it was believed that there were only two species of 'green' frog in Europe, and the Edible Frog was one of them. However it is now known that there are several species, and that the Edible Frog is the result of hybridisation between the Pool Frog (*Rana lessonae*) and the Marsh Frog (*R. ridibunda*). This hybridisation may have occurred as a result of habitat changes brought about by man. Variable in colour, the Edible Frog is usually green or blue-green above, with black blotches on its back, sides and legs, and it has a pale green or yellow vertebral stripe. The most aquatic of the green frogs, the Edible Frog is found in ponds, lakes and gravel pits, marshland and pools

| J | F | M | A | M | J |
| J | A | S | O | N | D |

Sand Lizard
Lacerta agilis

ID FACT FILE

TOTAL LENGTH:
Up to about
24cm

TAIL LENGTH:
Up to about
12cm

FIELD CHARACTERS:
Large-headed,
often with green
on sides

SIGNS:
Usually none

SOUNDS:
Heard rustling
as it hides

SIMILAR SPECIES:
Other small
lizards generally
have smaller
heads

It is one of the most widespread species of lizard in Europe but, in the north, its range has become fragmented, with many local extinctions, particularly in Britain and Scandinavia. In Britain, colonies can be found in the suth and south-west of England, and north near Liverpool and Stockport. Its colouring is very variable, but the males usually have bright green markings on the flanks during the breeding season. It is a terrestrial lizard found in mostly dry habitats, in the north of its range often in heaths and sand dunes. It often clambers in heather and other low vegetation in order to bask in the sun, and it feeds mostly on insects and spiders, but also occasionally preys upon young lizards.

SNAKES

| J | F | M | A | M | J |
| J | A | S | O | N | D |

Grass Snake
Natrix natrix

ID FACT FILE

TOTAL LENGTH:
Up to 2m,
usually less

FIELD CHARACTERS:
Olive green with
black barring;
often yellow
collar

SIGNS:
None

EGGS:
Up to 25 or
more, oval

SOUNDS:
Hisses

SIMILAR SPECIES:
Viperine Snake

The most widespread and often most abundant snake in Europe, the Grass Snake occurs in an extremely wide range of habitats, often close to human habitation. There is some variation in the colouring and markings and several subspecies have been described. The basic colour is olive green with blackish barring and usually a bright yellow collar. It is generally found in damp habitats, and is often very aquatic, feeding on fish and amphibians, although it also eats a wide range of other prey. It lays its eggs in piles of rotting vegetation (originally probably along riversides, but now often in manure heaps and compost heaps). With the decline of horses and other changes in farming, it has declined considerably.

SNAKES

J	F	M	A	M	J
J	A	S	O	N	D

Adder
Vipera berus

ID FACT FILE

TOTAL LENGTH:
Up to 90cm,
usually less
than 65cm

FIELD CHARACTERS:
Brown (female)
or grey (male),
with dark zig-zag

SIGNS:
None

EGGS:
Gives birth to
5–15 living young

SOUNDS:
Hisses

SIMILAR SPECIES:
Other vipers;
Viperine Snake

The most widespread snake in Europe and the only one to occur as far north as the Arctic Circle. Like other adders it gives birth to living young, and they are able to develop because of its ability to keep warm by basking. It occurs in a wide range of habitats, but in the northern parts of its range often on moors, heaths, sand dunes, whereas in the south it is mostly in mountains (at altitudes of up to 3,000m). It swims well, and is seen in water more commonly than other vipers. The males have an elaborate courtship 'dance'. Although venomous, its bite is rarely fatal to humans.

FISH

J	F	M	A	M	J
J	A	S	O	N	D

European Eel
Anguilla anguilla

ID FACT FILE

SIZE:
51 cm (male),
142 cm (female)

DESCRIPTION:
Elongated, 'eel-like'. Small pelvic fins near the head. Anal and dorsal fins begin further back and enclose the tail up to the tip. In fresh water back brownish, yellow below; migrating adults in sea black above, silver below

FOOD:
Invertebrates and small fish; larvae feed on plankton

LOOKALIKES
American Eel
A. rostrata.
Similar but fewer vertebrae; reported in Denmark

Adult eels travel to spawn in the south-west North Atlantic around the Sargasso Sea. Here the adults die, but millions of eggs are produced which hatch into transparent, flattened larvae, and are carried to the eastern Atlantic, taking about 12 months. They continue their journey north, now known as 'elvers', growing into yellow eels over a number of years. After a further period of growth males and females become silver eels.

FISH

J	F	M	A	M	J
J	A	S	O	N	D

Pike
Esox lucius

ID FACT FILE

Size:
Length 90cm (male), 150cm (female)
Weight 25kg

Description:
large freshwater fish with a long snout and jaws. Olive green above, sides show pale spots and bars; white or yellow below

Food:
fish; sometimes small mammals and young water birds

Lookalikes:
European mud-minnow *Umbra krameri* is a small relative of the pike with short jaws and a blunt snout

A large predator with a long snout and jaws and powerful teeth, distinctive patterns on sides are unique to larger specimens. Found in lakes, slow-moving sections of rivers, and brackish areas towards the north of its range. The pike hunts alone, among vegetation waiting motionless to pounce on prey. Spawning takes place in spring with each female courted by two or more males. Eggs are laid on vegetation in shallow water up to 500,000 tiny, yellowish in clusters of oil droplets. They hatch after 10–15 days, and fish are mature by 2–4 years when they will be 18–30 cm. Young pike may feed on invertebrates, but the usual food for adult pike is fish.

FISH

Three-spined Stickleback

Gasterosteus aculeatus

J	F	M	A	M	J
J	A	S	O	N	D

ID FACT FILE

SIZE:
Length 11.5 cm

DESCRIPTION:
Blue-green with a silver tinge above, white below. Breeding males have bright red belly and underside of head

FOOD
plant material

LOOKALIKES
Nine-spined Stickleback *Pungitius pungitius*, found in fresh and brackish waters. Fifteen-spined Stickleback *Spinachia spinachia*, a marine stickleback with an elongated body and slender tail

These small fish, with their characteristic three spines appearing before the dorsal fin, are the most common small freshwater species in Britain. Different forms of the species are separated by the extent of lateral bony plates: trachurus has a complete series from the pelvic fin to the caudal fin on the tail; *G. semiarmatus* has a partial set with a short series of plates; *G. leiurus* has plates only on the sides of the abdomen; in *G. gymnurus* the plates are completely absent. Breeds in fresh water, eggs are guarded by the male.

FISH

| J | F | M | A | M | J |
| J | A | S | O | N | D |

Roach

Rutilus rutilus

ID FACT FILE

SIZE:
Length 50cm
Weight 2kg

DESCRIPTION:
Dark, bluish
green above,
sides golden or
brassy, silver-
white under-
neath, fins
orange-red or
darker.

FOOD:
Feeds on bottom
on invertebrates,
especially mol-
luscs, and also
plant material

LOOKALIKES:
Rudd *Scardinus
erythropalmus*
has a browner
body, sometimes
with a gold tinge;
origin of dorsal
fin is well behind
the base of the
pelvic fin

The Roach is a frequently-caught coarse fish,
which has the origin of its dorsal fin directly
above the base of the pelvic fin. Deep-bodied,
the Roach is found in schools in lakes and
slow-moving parts of rivers, and sometimes in
brackish water. The Roach breeds in late
spring to early summer, in fresh water, laying
5,000–200,000 tiny, greenish eggs usually in
shallow water, over plants and sometimes gravel. Fish matures at 2–3 years (male) and 3–4
years (female). Hybridises with Rudd, Bleak,
Chub, Bream and Silver Bream. Related
species are found in Europe, including the
Danubian Roach, Black Sea Roach, as well as
an Alpine subspecies.

FISH

J	F	M	A	M	J
J	A	S	O	N	D

Perch
Perca fluviatilis

ID FACT FILE

Size:
Length 50cm
Weight 4.75kg

Description
Dark green above, lateral lines on sides, lighter below. Two prominent dorsal fins very close together; dark blotch at base of first dorsal fin. Pelvic and anal fins reddish

Food
fish and invertebrates

Lookalikes
Ruffe *Gymnocephalus cernuus* with which it hybridizes

This popular sport fish has characteristic large fins and striped sides. It inhabits lakes and slow rivers, and often shelters near roots or other vegetation. It spawns in late April–May in water 0.5–3m deep, and females are accompanied by several males. Pale yellowish eggs are laid over weed, up to 200,000. They appear as long white strands. The eggs hatch when larvae are 4.8–6.4mm; fish mature by 3–4 years. The Perch hybridizes with the smaller Ruffe (15cm), which is dark green with black spots on its back and sides, and yellowish below. The Ruffe inhabits the bottom of slow rivers and lakes in small schools.

FISH

J	F	M	A	M	J
J	A	S	O	N	D

Bream
Abramis brama

ID FACT FILE

SIZE:
Length 80 cm
Weight 11.5 kg

DESCRIPTION:
Deep, flattened body, grey-brown above, silvery sides and below, larger fish may have a bronze tinge. Thick lips

FOOD:
Invertebrates, such as insect larva

LOOKALIKES:
Silver Bream *Blicca bjoerkna*, Roach *Rutilus rutilus*, Rudd *Scardinus erythropalmus*

This large, deep-bodied fish is found in slow-moving rivers and brackish waters in lowland areas. During the winter, schools of bream move to deeper water. It feeds on invertebreates sucked from mud on the river bottom forming 'bream pits'. Breeds April–June; spawns at night in shallow waters over vegetation such as weeds. Up to 340,000 yellowish eggs are laid and these hatch after about 2 weeks when the larvae are 4.6–5.3 mm. Fish mature at 23 cm (male), 21 cm (female), at about 5 years. The Bream hybridizes with Roach, Rudd and Silver Bream.

FISH

Carp
Cyprinus carpio

ID FACT FILE

SIZE:
Length 103 cm
Weight 34 kg

DESCRIPTION:
Deep-bodied,
elongated fish
with a long dor-
sal fin and
promient pec-
toral, pelvic and
anal fins, all the
same colour as
the body which is
greenish-brown
above, becoming
yellowish-brown
below

FOOD:
Omnivorous

LOOKALIKES:
Many varieties
and hybrids
which are
cultivated such
as the colourful
Koi, Mirror and
Leather Carp

Orignally introduced from Asia. *C. carpio* is immediately distinguished from its relatives the goldfish and Crucian Carp by four (two long, two short) barbels or sensory spines on the mouth and jaw. The carp lives in slow, stagnant or muddy parts of lakes and rivers that are well-vegetated. Spawns in late spring to summer; eggs are laid over weed or grass in shallow water. Females are attended by several males. Huge numbers of eggs are laid which hatch in about a week when larvae are 5 mm in length. Fish are sexually mature by 5 years.

INDEX

COMMON NAMES

LATIN NAMES